☞ What's New in This Edition ☜

This book was first published in 2000 and this is the first major revision. The main changes are:

1. The "Advanced Systems" chapter *(pp. 86–89)* —formerly called "Alternative Systems"— has been completely rewritten. In recent years, many devices have become available that are utilized when a conventional gravity-powered septic system is deemed insufficient to provide adequate wastewater treatment. We have concluded that mounds are no longer the state of the art in advanced technology. There are now simpler, smaller, less environmentally destructive systems that treat wastewater better than mounds. (What's ironic here is that many areas in the country are just beginning to mandate mounds.) We cover recirculating sand filters, the new "trickling biofilters," effluent filters, and drip irrigation, three different devices for shallow drainfields, as well as two systems for restoring failed drainfields. There are numerous website references for further information on advanced systems.
2. Two new chapters, "Excessive Engineering and Regulatory Overkill" *(pp. 91–110)*, and "Tale of Two Sewers" by John Hulls *(pp. 111–116)* describe an unfortunate turn of events over the last decade or so in North America. Engineers and regulators have coordinated efforts in various parts of the country to force homeowners into unnecessarily high-tech, expensive septic systems. Much of this is being done with no scientific proof of septic pollution and does not even adhere to national standards as prescribed by the Environmental Protection Agency. Community-wide plans are often designed to be as expensive as possible in order to obtain maximum "clean water" grant funding. Individuals are forced to install environmentally destructive mounds in soils where simple gravity-powered septic systems have worked for many decades.

   We describe several case studies of towns confronted by poorly engineered, overly expensive plans devised by consortiums of engineers, regulators, and special interests. We tell homeowners what to watch out for and how to deal with this new onsite wastewater *modus operandi* when it appears. We also encourage you to let us know your experiences in such situations so we can continue documenting what is unfortunately becoming a national trend.
3. We describe the great importance of DNA testing to determine if pollution is human or animal in origin *(p. 110)*. Many multi-million dollar plans have been based on faulty science.
4. We have updated and added many new websites for obtaining further information on many aspects of onsite wastewater treatment and disposal.

The heart of this book is still the gravity-powered septic system, an elegant, environmentally sound, economical, effective method of dealing with wastewater. No pumps, no motors, with harmful organisms neutralized by natural soil organisms. At the same time we believe that knowing the characteristics of today's advanced devices can help homeowners in making intelligent decisions and in interacting in an informed manner with engineers and health officials.

Septic Bulletin Board: We will update readers with new information on our website:

www.shelterpub.com/_septic/bulletin_board.html

We invite you to contact us with information you think will help other homeowners.

HERE'S WHAT REVIEWERS HAVE SAID ABOUT
THE SEPTIC SYSTEM OWNER'S MANUAL

"This book is a long-overdue guide to how and why these systems work—or don't . . . an entertaining discussion of a highly technical, but elegantly simple, process."

–*Fine Homebuilding*

"This is a great book."

–*Home Power* Magazine

". . . a new book dedicated to increasing mankind's knowledge of the gravity-powered systems for human waste disposal."

–*Fresno Bee*

". . . we are delighted by this book's excellent illustrations… and clear explanations. Here's a straightforward easy-to-understand book on small-scale wastewater disposal."

–*Builders Booksource Newsletter*

". . . a do-it-yourself guide to the natural process that takes place underground and the simple technology needed to make it work."

–*Small Flows Journal*

". . . well-written, easy to read, and which contains about all the information any owner could want."

–*Ecology Action Newsletter*

"It's all here: system mechanics, the biology, maintenance, do's and don'ts . . ."

–*Point Reyes Light*

"Any homeowner with a septic tank should get a copy."

–*Nashville Tennessean*

". . . an excellent guide essential for any who reside on a septic system."

–*Wisconsin Bookwatch*

"Even homeowners whose relationship with their septic tank is limited to calling the plumber will find valuable, cost-saving advice here . . ."

–*Booklist*

"APPROXIMATELY ONE FOURTH (26 MILLION) OF THE ESTIMATED 124 MILLION HOUSING UNITS IN THE UNITED STATES ARE SERVED WITH SEPTIC TANKS OR CESSPOOLS, ACCORDING TO THE 2005 AMERICAN HOUSING SURVEY PREPARED BY THE U. S. CENSUS BUREAU."
HTTP://WWW.CENSUS.GOV/HHES/WWW/HOUSING/AHS/NATIONALDATA.HTML

This revised edition is dedicated to the memory of Peter Aschwanden, artist extraordinaire.

REVISED · UPDATED

BY LLOYD KAHN
ILLUSTRATED BY PETER ASCHWANDEN
CONTRIBUTING EDITOR, JOHN HULLS

SHELTER PUBLICATIONS
BOLINAS, CALIFORNIA, U.S.A.
WWW.SHELTERPUB.COM

Distributed in the United States and Canada by Publishers Group West

Library of Congress Cataloging-in-Publication Data

The Library of Congress catalogued the first edition as follows:

Lloyd Kahn, 1935–
Septic system owner's manual / Lloyd Kahn, Blair Allen & Julie Jones ; illustrations by Peter Aschwanden.
p. cm.
Includes bibliographical references and index.
ISBN: 0-936070-20-X
1. Septic tanks — Popular works. I. Allen, Blair. II. Jones, Julie, 1943 Mar. 24– III. Title.
TD778 .B58 1999
628'.742–dc21

98-28772
CIP

ISBN Numbers for *The Septic System Owner's Manual, Revised Edition*
ISBN-10: 0-936070-40-4
ISBN-13: 978-0-936070-40-7

We are grateful to the following publisher for permission to reprint portions of previously published material:

Stein and Day: First Scarborough Books Edition, New York, for drawings of microbes on pp. 34–35 from *Microbe Power — Tomorrow's Revolution*, by Brian J. Ford, © 1978.

12 11 10 9 8 7 — 11 10 09 08 07
(Lowest digits indicate number and year of latest printing.)

Printed in the United States of America

Additional copies of this book may be purchased at your favorite bookstore or by sending $17.95 plus $5.00 shipping charges to:

Shelter Publications, Inc.
P. O. Box 279
Bolinas, California 94924
415-868-0280
Orders, toll-free: 1-800-307-0131
Email: shelter@shelterpub.com

Visit Our Website
SHELTER ONLINE
http://www.shelterpub.com

	What It's About	viii
CHAPTER 1	The Tank	2
CHAPTER 2	The Drainfield	12
CHAPTER 3	The Soil	28
CHAPTER 4	Down the Drain	36
CHAPTER 5	Septic System Maintenance	46
CHAPTER 6	Red Alert!	55
CHAPTER 7	Graywater Systems	62
CHAPTER 8	Composting Toilet Systems	72
CHAPTER 9	Advanced Systems	84
CHAPTER 10	Excessive Engineering and Regulatory Overkill	100
CHAPTER 11	A Tale of Two Sewers	111
CHAPTER 12	Small Town Septic System Upgrades	117
CHAPTER 13	A Brief History of Wastewater Disposal	133
	Appendix	156
	Index	176

AN UNHERALDED WONDER

The gravity-powered septic system is a wonder of technology — past and present. Its operation is so quiet, natural, and energy-free that we tend to forget the vital function it serves.

Sewage is carried from the house to the tank via gravity — no motors, no fossil-fuel energy consumption, no noise. Wastewater goes from the tank to the drainfield — also via gravity — where microorganisms in the soil digest and purify bacteria and viruses. When the soil is suitable and the system healthy, it is an example of efficient design and natural forces, returning clean water to the water table (or to plants or the air) — all functioning silently under the surface of the earth.

WHAT IS THIS BOOK?

There are currently more than 25 million septic systems in the United States. Moreover, each year some 400,000 new systems are built. Yet in spite of such widespread usage, the average homeowner seems to know little about the basic operation and appropriate maintenance of a septic system.

This book describes the conventional gravity-fed septic system, how it works, how it should be treated (what should and should not go down the drain), how it should be maintained, and what to do if things go wrong. There is also basic information on the recent evolution in composting toilet systems, designs for simple graywater systems, and some of the typical alternatives to the standard, gravity-fed septic system.

There is a chapter with advice to any community faced with town-wide septic upgrades, and last, an illustrated chapter on the history of waterborne waste disposal.

As you will see, this is not an engineering treatise. Nor do we cover any of the many nonconventional systems in use in various parts of the country by a variety of wastewater engineers and soil scientists. This is a basic manual for the average homeowner, based on conventional systems, providing practical advice on how to keep these systems up (or should we say down?) and running.

WHO IS THIS BOOK FOR?

Primarily homeowners (or home dwellers), but also for builders, architects, plumbers, septic contractors, pumpers, and realtors, as well as health departments, wastewater districts, and small towns—anyone who wants to understand these very important, but often misunderstood, working principles.

If you are buying a house with a septic system, it is very important that you understand septic basics, so that you know what you are getting.

WHY THE NEED FOR THIS BOOK?

Homeowners will find this book useful in terms of:

- *Working systems:* By understanding septic system principles, you will know how to treat your system intelligently and maximize its useful life.
- *Partially failing systems:* By changing daily household practices, and perhaps making minor repairs, you may be able to nurse along an ailing system or even bring it back to life.
- *Failing systems:* You will be given a discovery process to search for the problem in a given order. You will discover if the problem is relatively easy to fix (as with pipe blockage), or major (drainfield failure). You will understand what went wrong and be given a variety of options for repair.
- *Alternative systems:* By this we mean alternatives to the gravity-powered system—typically, mounds, pressure-dosed drainfields, sand filters, etc.—often required by health officials these days. You will be given the basics of these designs so you will understand how they work and what purpose they serve.

Note to wastewater districts, health departments, and other regulatory agencies:

See p. 180 on obtaining copies of *Homeowner's Septic System Guide* for homeowners in your district.

SOME CAVEATS

The local angle: Although the principles described here are more or less the same all over the world, there are local factors of soil and climate, as well as practical experience, that will differ from region to region. Once you understand the basics, we suggest you talk to local builders, septic tank pumpers, and homeowners. There is no substitute for local experience.

The comprehensive angle: We do not cover everything on the subject. We do not describe all possible systems in all parts of the world. Our intent here is to give you the basics, so you can make informed judgments on maintenance, repairs, and upgrades.

The appropriate technology angle: Unfortunately, regulatory agencies have tended to require higher-tech, more expensive systems in recent years. In some cases, this approach is necessary, but many times it's overkill. Granted that there will be situations where soil and/or climate require other options, yet the gravity-fed system remains the simplest and most ecological design; it is the "stick shift" of septic systems, and therefore, the heart of this book.

The varying opinions angle: Experts in the field all have different opinions. We have consulted a number of professionals and have attempted to strike a balance as to sensible and useful information for homeowners.

ONGOING SEPTIC INFO

We will update, supplement, and correct this book in future editions and on the web. Contact us at Shelter Publications, P. O. Box 279, Bolinas, CA 94924, by fax at 415-868-9053, or online at septic@shelterpub.com if you have anything to contribute. We welcome corrections, additions, and insights, and will post useful information you wish to share on our website.

1
2
3
4
5
6
① DRAIN FROM HOUSE
② TWO-COMPARTMENT SEPTIC TANK
③ DRAINFIELD
④ EFFLUENT DISTRIBUTION PIPES
⑤ SOIL WHERE EFFLUENT IS PURIFIED
⑥ WATER TABLE

IN A NUTSHELL . . .

A TYPICAL CONVENTIONAL SEPTIC SYSTEM CONSISTS OF TWO BASIC WORKING PARTS:

- A SEPTIC TANK
- AN UNDERGROUND DISPOSAL FIELD (VARIOUSLY CALLED DRAINFIELD, LEACHFIELD, SOIL ABSORPTION FIELD)

WASTEWATER FLOWS FROM THE HOUSE TO THE TANK.

EFFLUENT (THE WASTEWATER MINUS SOLIDS) FLOWS FROM THE TANK TO THE DRAINFIELD.

THE MAIN JOB OF THE TANK IS TO INTERCEPT THE SOLIDS.

THE MAIN JOB OF THE DRAINFIELD IS TO PURIFY AND DISPERSE THE EFFLUENT.

THE SOIL FILTERS AND PURIFIES THE EFFLUENT, WHICH THEN RETURNS TO THE WATER TABLE, IS TAKEN UP BY PLANTS, AND/OR EVAPORATES.

IN THE DRAWING AT THE LEFT, YOU CAN SEE HOW ALL THIS WORKS FOR A TYPICAL HOUSE.

THIS BOOK IS A MORE DETAILED DESCRIPTION OF THIS PROCESS.

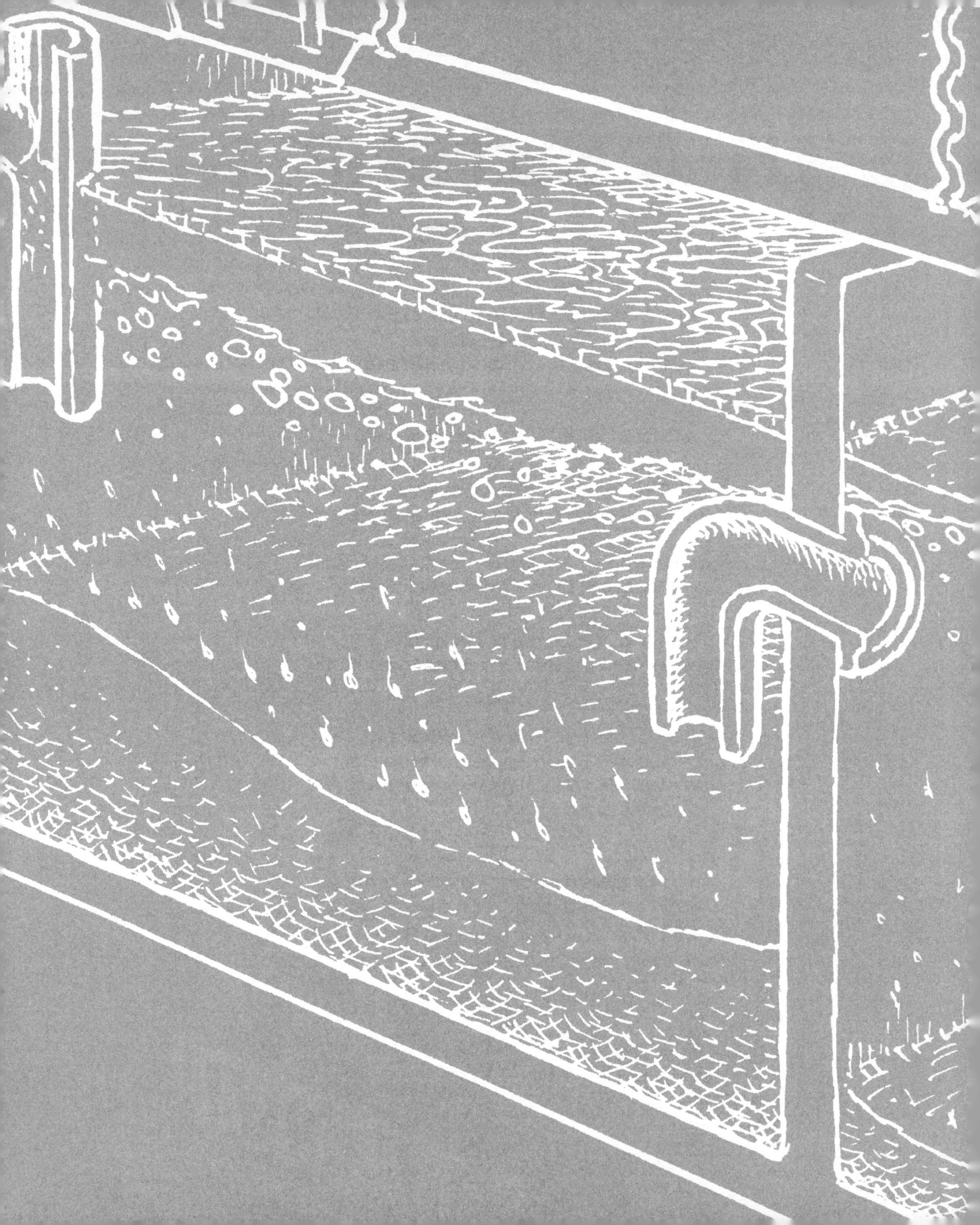

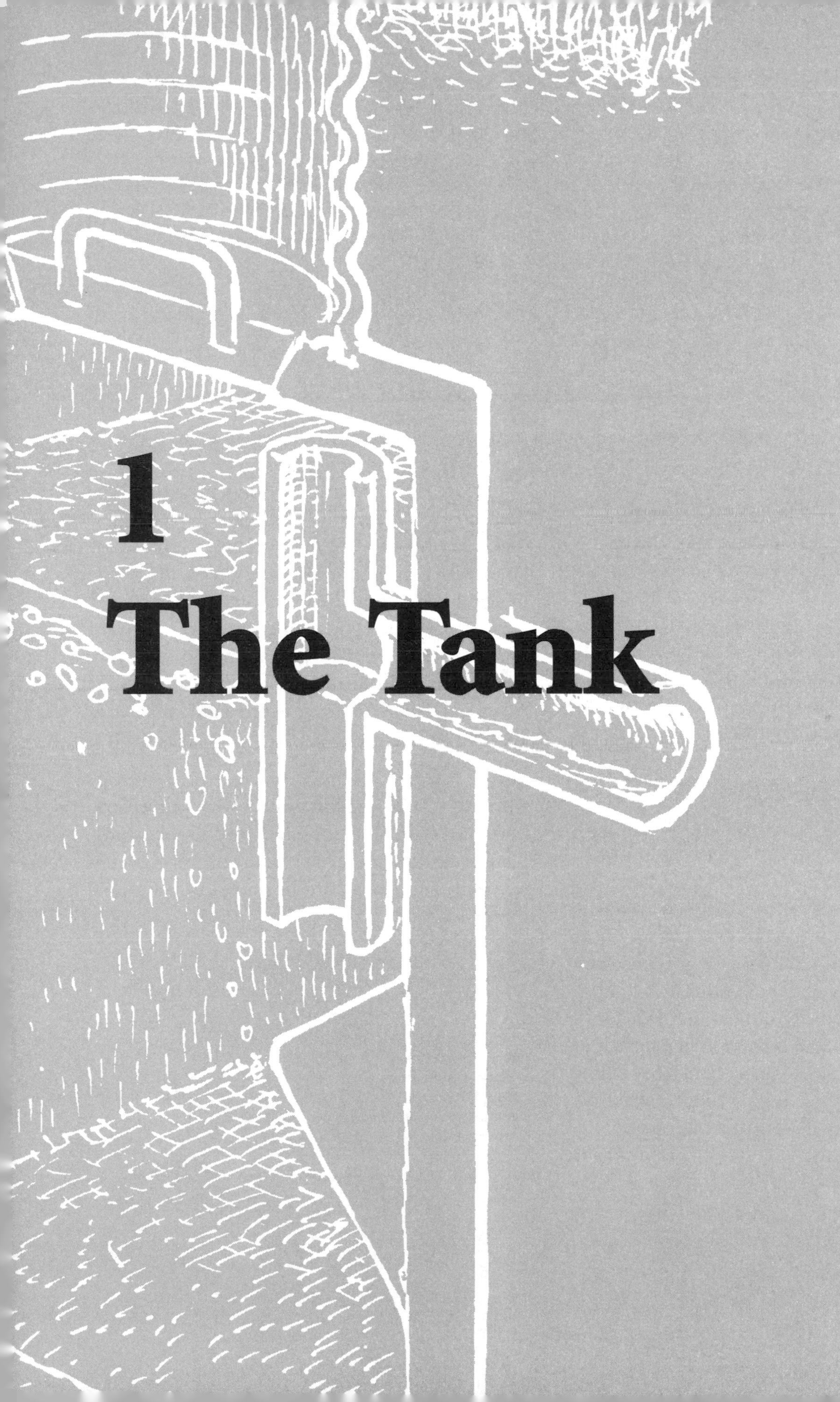

1
The Tank

CHAPTER ONE

The Tank

Everything that goes down any of the drains in the house (toilets, showers, sinks, laundry machines) travels first to the septic tank. The septic tank is a large-volume, watertight tank which provides initial treatment of the household wastewater by intercepting solids and settleable organic matter before disposal of the wastewater (effluent) to the drainfield.

SEPTIC TANK CONSTRUCTION

Early tanks were often redwood (with limited life spans) and later, polyethylene (many with structural difficulties). Modern septic tanks are usually made of concrete or fiberglass or plastic and are available as prefabricated units. (Fiberglass and plastic tanks are lightweight and often used at sites where a truck cannot get in to deliver a concrete tank.) Many homemade tanks have been constructed of concrete block, but they are difficult to make watertight. (See below on the importance of a tank being watertight.) Most tanks are and will continue to be concrete. *(See p. 58 for more complete information on the different types of tanks.)*

SEPTIC TANK SIZE

For single-family homes, tanks typically range in size from 500 to 1500 gallons of wastewater storage capacity. For a one- or two-bedroom home, a 1000- or 1200-gallon tank is common; for a three-bedroom home, a 1500-gallon tank. A typical 1000-gallon concrete tank measures about nine feet long by five feet wide and five feet deep.

Although a 1500-gallon tank costs slightly more than a 1000-gallon one, it affords more complete digestion and can reduce pumping occurrences by a factor of four or more for a family of three.

THE MAIN JOB OF THE SEPTIC TANK IS TO INTERCEPT THE SOLIDS. THE TANK SEPARATES THE "FLOATERS" AND "SINKERS" FROM THE LIQUID, OR EFFLUENT, WHICH FLOWS OUT TO THE DRAINFIELD.

OVERALL PLACEMENT AND CONFIGURATION OF TANK

The septic tank is located near the house and is buried with the top of the tank about a foot or two below the surface of the ground. There is an inlet port and an outlet port through the sidewalls on opposite ends of the tank for wastewater flow. The interior may be a single open chamber, but commonly consists of two compartments created by an internal wall with an opening for flow from one compartment to the next.

MANHOLES & RISERS

The top of the tank has covered, removable manholes to allow for routine inspection and pumping out of the interior. For easier location and access to the manholes, a riser can be constructed over each manhole, extending from the tank top to the ground surface. *(See p. 50 for a prefabricated riser that can be retrofitted onto an existing tank.)*

BUILD YOUR OWN?

Plans for building your own tank with concrete block or cast-in-place concrete appear in *Wells and Septic Systems,* by Max and Charlotte Alth. *(See p. 168.)*

COLD CLIMATES

In colder climates, the tank may be buried deeper below the frostline to avoid damage from freezing. Also, foam insulation (such as Dow Blueboard, a burial-rated, 2-inch insulation) can be placed over the top and along the sides of the tank to prevent freezing. In some parts of Canada, a light bulb inside the tank is used for heat.

ONE-COMPARTMENT VS. TWO-COMPARTMENT TANKS

The two-compartment septic tank has remained almost unchanged for the last 100 years. The initial reason for two compartments was to provide a structural wall to support the top of the tank. Most installed tanks these days are of the two-compartment variety, with a vertical partition dividing the tank into ⅔ to ⅓ sections. However, some wastewater engineers now think that the single compartment tank performs as well as or better than the two-compartment tank. *(See p. 9.)*

FUNCTION OF THE SEPTIC TANK

While relatively simple in construction and operation, the septic tank provides a number of important functions through a complex interaction of physical and biological processes. The essential functions of the septic tank are to:

- receive all wastewater from the house
- separate solids from the wastewater flow
- cause reduction and decomposition of accumulated solids
- provide storage for the separated solids (sludge and scum)
- pass the clarified wastewater (effluent) out to the drainfield for final treatment and disposal

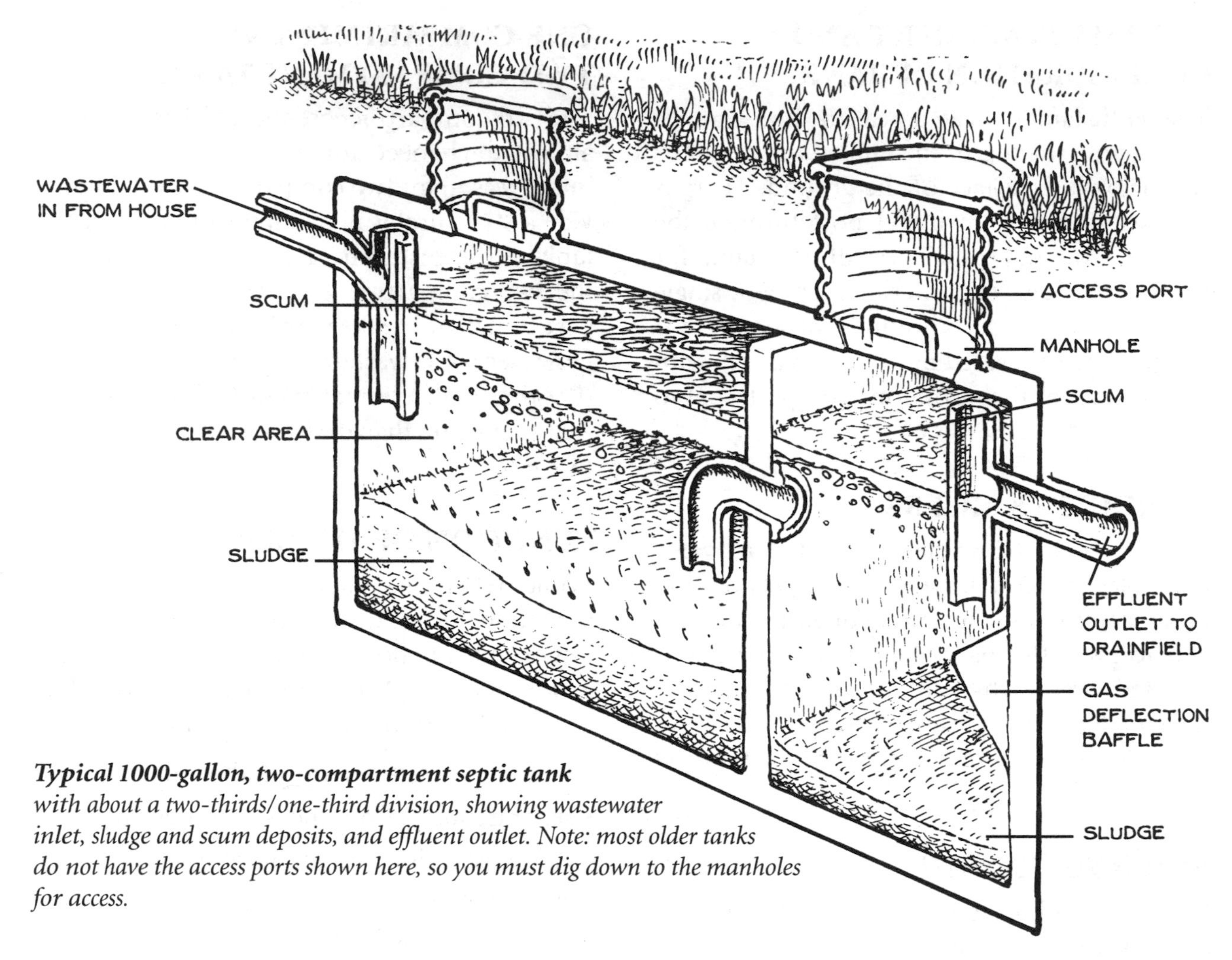

Typical 1000-gallon, two-compartment septic tank *with about a two-thirds/one-third division, showing wastewater inlet, sludge and scum deposits, and effluent outlet. Note: most older tanks do not have the access ports shown here, so you must dig down to the manholes for access.*

Primary Treatment

As stated, the main function of the septic tank is to remove solids from the wastewater and provide a clarified effluent for disposal to the drainfield. The septic tank provides a relatively quiescent body of water where the wastewater is retained long enough to let the solids separate by both settling and flotation. This process is often called *primary treatment* and results in three products: *scum, sludge,* and *effluent.*

Scum: Substances lighter than water (oil, grease, fats) float to the top, where they form a scum layer. This scum layer floats on top of the water surface in the tank, Aerobic bacteria work at digesting floating solids.

Sludge: The "sinkable" solids (soil, grit, bones, unconsumed food particles) settle to the bottom of the tank and form a sludge layer. The sludge is denser than water and fluid in nature, so it forms a flat layer along the tank bottom. Underwater anaerobic bacteria consume organic materials in the sludge, giving off gases in the process and then, as they die off, become part of the sludge.

Effluent: Effluent is the clarified wastewater left over after the scum has floated to the top and the sludge has settled to the bottom. It is the clarified liquid between scum and sludge. It flows through the septic tank outlet into the drainfield.

HOW LONG LIQUIDS MUST REMAIN IN TANK

Effective volume: The floating scum layer on top and the sludge layer on the bottom take up a certain amount of the total volume in the tank. The *effective volume* is the liquid volume in the clear space between the scum and sludge layers. This is where the active solids separation occurs as the wastewater sits in the tank.

Retention time: In order for adequate separation of solids to occur, the wastewater needs to sit long enough in the quiescent conditions of the tank. The time the water spends in the tank, on its way from inlet to outlet, is known as the *retention time.* The retention time is a function of the effective volume and the daily household wastewater flow rate:

$$\text{Retention Time (days)} = \frac{\text{Effective Volume (gallons)}}{\text{Flow Rate (gallons per day)}}$$

A common design rule is for a tank to provide a *minimum* retention time of at least 24 hours, during which one-half to two-thirds of the tank volume is taken up by sludge and scum storage. Note that this is a *minimum* retention time, under conditions with a lot of accumulated solids in the tank. Under ordinary conditions (i.e., with routine maintenance pumping) a tank should be able to provide two to three days of retention time.

As sludge and scum accumulate and take up more volume in the tank, the effective volume is gradually reduced, which results in a reduced retention time. If this process continues unchecked—if the accumulated solids are not cleaned out (pumped) often enough—wastewater will not spend enough time in the tank for adequate separation of solids, and solids may flow out of the tank with the effluent into the drainfield. This can result in clogged pipes and gravel in the drainfield, one of the most common causes of septic system failure. *(See pp. 58–59 for more about this.)*

SOLIDS STORAGE

In order to avoid frequent removal of accumulated solids, the septic tank is (hopefully) designed with ample volume so that sludge and scum can be stored in the tank for an extended period of time. A general design rule is that one-half to two-thirds of the tank volume is reserved for sludge and scum accumulation. A properly designed and used septic system should have the capacity to store solids for about five years or more. However, the rate of solids accumulation varies greatly from one household to another, and actual storage time can only be determined by routine septic tank inspections. *(See pp. 49–52.)*

ANAEROBIC DECOMPOSITION

While fresh solids are continually added to the scum and sludge layers, anaerobic bacteria (bacteria that live without oxygen) consume the organic material in the solids. The by-products of this decomposition are soluble compounds, which are carried away in the liquid effluent, and various gases, which are vented out of the tank via the inlet pipe that ties into the house plumbing air vent system.

THE MOST SERIOUS PROBLEM WITH SEPTIC SYSTEMS OCCURS WHEN SLUDGE AND/OR SCUM ARE CARRIED OUT INTO THE DRAINFIELD. THIS CAN CAUSE SURFACING OF EFFLUENT OR SYSTEM BACKUP INTO HOUSE DRAINS.

Anaerobic decomposition results in a slow reduction of the volume of accumulated solids in the septic tank. This occurs primarily in the sludge layer but also, to a lesser degree, in the scum layer. The volume of the sludge layer is also reduced by compaction of the older, underlying sludge. While a certain amount of volume reduction occurs over time, sludge and scum layers gradually build up in the tank and eventually must be pumped out.

FLOW INTO AND OUT OF THE TANK

The inlet and outlet ports of the tank are generally equipped with devices such as baffles, concrete tees, or in more recent years, sanitary tees (T-shaped pipes with one short and one long leg).

Inlets

The inlet device dissipates the energy of the incoming flow and deflects it downwards. The vertical leg of the tee extends below the liquid surface well into the clear space below the scum layer. This prevents disturbance of the floating scum layer and reduces disruptive turbulence caused by incoming flows. The inlet device also is supposed to prevent short-circuiting of flows across the water surface directly to the outlet.

The upper leg of the inlet should extend well above the liquid surface in order to prevent floating scum from backing up into, and possibly plugging, the main inlet pipe. The open top of the inlet tee allows venting of gases out of the tank through the inlet pipe and fresh air vents of the household plumbing.

Outlets

The outlet device is designed to retain the scum layer within the tank. A sanitary tee can be used with the lower leg extending below the scum layer. The elevation of the outlet port should be 2 to 3 inches below the elevation of the inlet port. This prevents backwater and stranding of solids in the main inlet pipe during momentary rises in the tank liquid level caused by surges of incoming wastewater.

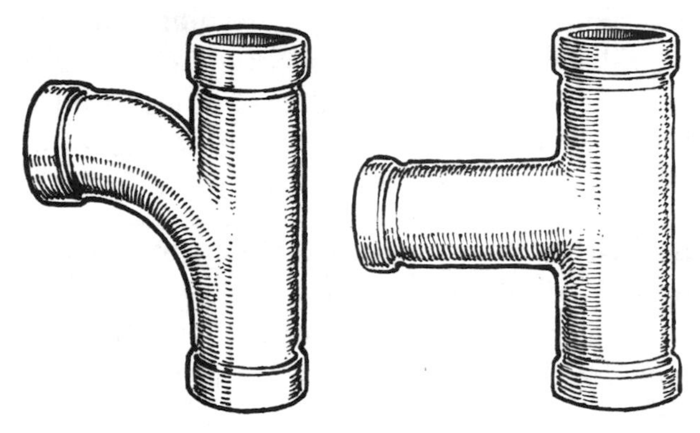

Typical inlet/outlet tees

Gas Deflection Baffle

Gases are produced by the natural digestion of sludge at the bottom of the tank, and particles of sludge can be carried upward by these rising gases. Some tanks have a gas deflection baffle, which prevents gas bubbles (to which solid particles often adhere) from leaving the tank by deflecting them away from the outlet and preventing them from entering the drainfield. *(See drawing, p. 6.)*

THE EFFLUENT FILTER

In newer systems, there is often an *effluent filter,* a simple device that, if properly maintained, will prolong the life of the drainfield. They range from 4 to 18 inches in diameter. *(See opposite page.)* As we have described, the most serious problem with septic systems is the migration of solids, grease, or oil into the drainfield, and the filter is effective in preventing this.

A filter restricts and limits passage of suspended solids into the effluent. Solids in a filtered system's effluent discharge are significantly less than those produced in a non-screened system.

Moreover, the filter is relatively inexpensive (under $200) and can be quickly installed (retrofitted) in older tanks. The filter cartridge is removed and hosed off. The filters shown on this page are manufactured by in Oregon by Orenco Systems, Inc., 541-459-4449. Another manufacturer is Zabel Environmental Products, of Louisville, KY. *(See p. 174.)*

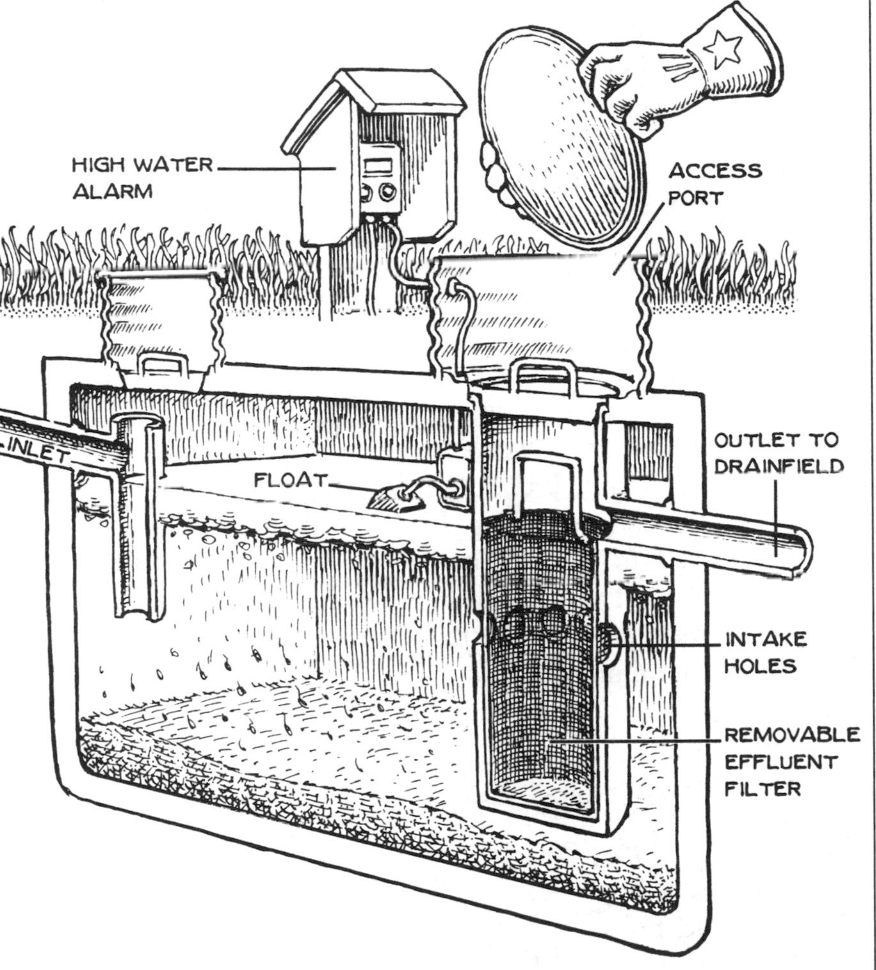

One-compartment septic tank with effluent filter
The filter is pulled out by hand periodically and hosed off. This relatively new accessory—when utilized regularly—can keep solids out of the drainfield.

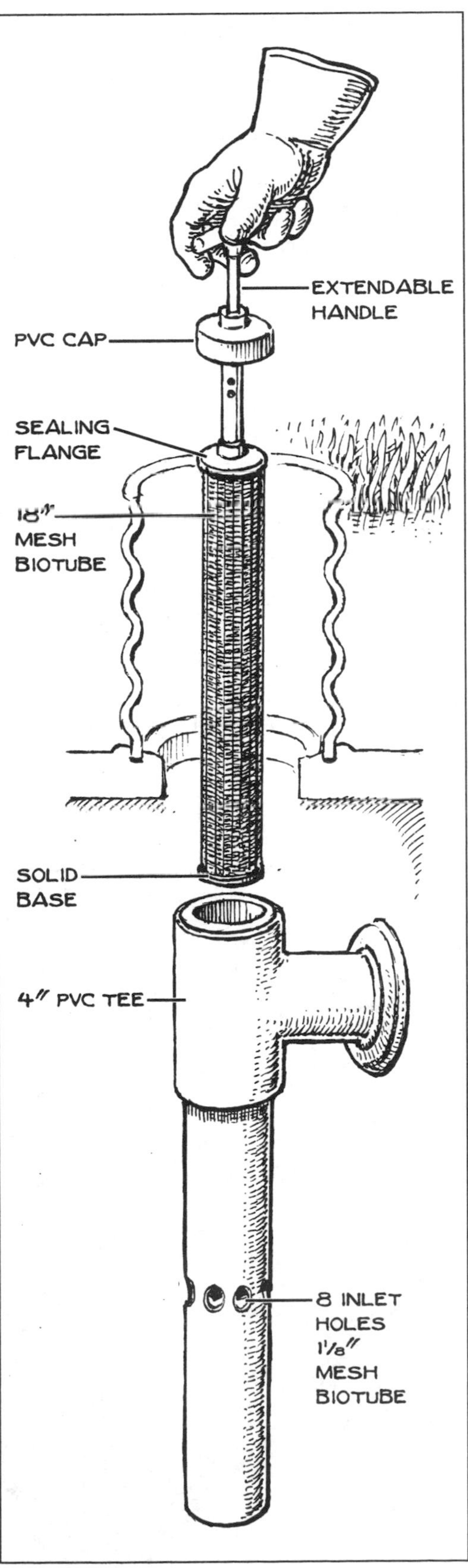

Retrofit effluent filter by Orenco
This filter—narrower than the one in the tank above—was designed so it could be retrofitted to an existing tank. It is easy to pull out and hose off.

FLOW BUFFERING

The septic tank also provides a buffering of flows between the house and the drainfield. Large surges from the household, such as toilet flushing or washing machine drainage, are dampered by the septic tank so that the flows leaving the tank and entering the drainfield are at substantially lower flow rates and extend over a longer period of time than the incoming surges.

THE MEANDER TANK

me·an·der *intr. v.* 1. To follow a winding and turning course. *n.* 1. A circuitous journey . . .

As explained on page 7, *retention time* (how long water remains in the tank before exiting to the drainfield) is important in the functioning of the system. Dr. John H. Timothy Winneberger, a well-respected innovator in sewage disposal technology, invented a triple-compartment tank —which Peter Warshall (in *Septic Tank Practices)* named the *meander tank.*

The meander tank actually increases the velocity of liquid, but smooths out the flow and helps stop short-circuiting from the inlet to the outlet. Two partitions are built as shown, dividing the tank into three rectangular chambers. Passage from one compartment to another is through slots too large to clog, and liquids traverse the tank's length three times, passing through one chamber, then the next two, before exiting to the drainfield. In the early model shown, the first chamber held about ⅔ of the liquid volume, the last about 1/9 of the liquid volume.

Note: Two-compartment meander tanks—easier to construct—are also built, with one longitudinal baffle.

The meander tank is considered by many engineers to be far more effective than conventional tanks, but the industry has not caught up with the vision. The meander tank is a better mousetrap, but no one is building it, at least not for homes. For commercial applications, engineer Michael D. Mitchell, of Northwest Septic, Inc., of Mt. Vernon, WA, has designed some very interesting meander tank configurations, intended, he says, to "minimize mixing while maximizing settling."

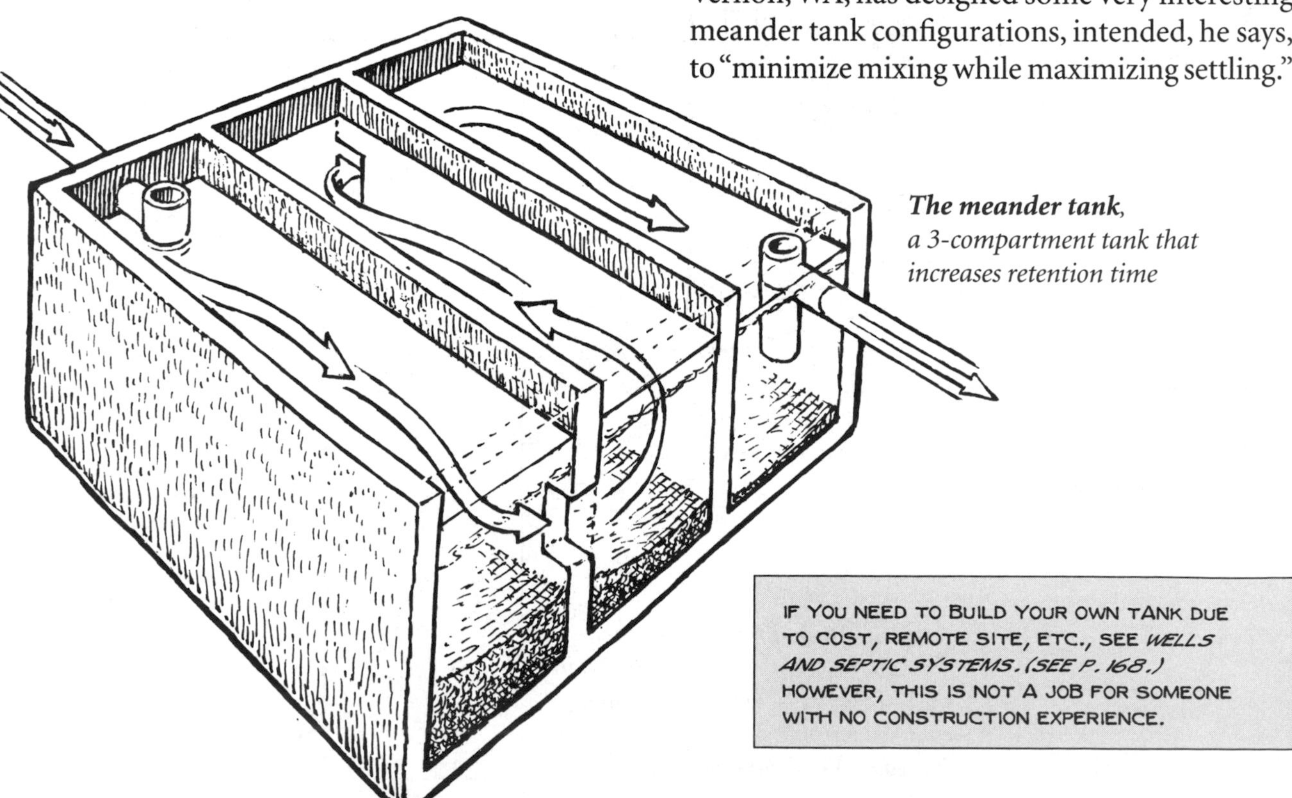

***The meander tank**, a 3-compartment tank that increases retention time*

IF YOU NEED TO BUILD YOUR OWN TANK DUE TO COST, REMOTE SITE, ETC., SEE *WELLS AND SEPTIC SYSTEMS*. (SEE P. 168.) HOWEVER, THIS IS NOT A JOB FOR SOMEONE WITH NO CONSTRUCTION EXPERIENCE.

MICROBES IN SEPTIC TANKS DIGEST, DISSOLVE, AND GASIFY COMPLEX ORGANIC WASTES

In 1907, W. P. Dunbar conducted tests on the decomposition of vegetable and animal matter in septic tanks. He stated, "The author has investigated the subject by suspending in septic tanks a large number of solid organic substances, such as cooked vegetables, cabbages, turnips, potatoes, peas, beans, bread, various forms of cellulose, flesh in the form of dead bodies of animals, skinned and unskinned, various kinds of fat, bones, cartilage, etc., and has shown that many of these substances are almost completely dissolved in from three to four weeks. They first presented a swollen appearance, and increased in weight. The turnips had holes on the surface, which gradually became deeper. The edges of the cabbage leaves looked as though they had been bitten, and similar signs of decomposition were visible in the case of other substances. Of the skinned animals, the skeleton alone remained after a short time; with the unskinned animals the process lasted rather longer. At this stage I will only point out that the experiments were so arranged that no portion of the substances could be washed away; their disappearance was therefore due to solution and gasification."

–*Principles of Sewage Treatment*, W. P. Dunbar, Charles Griffin & Co., London, 1907

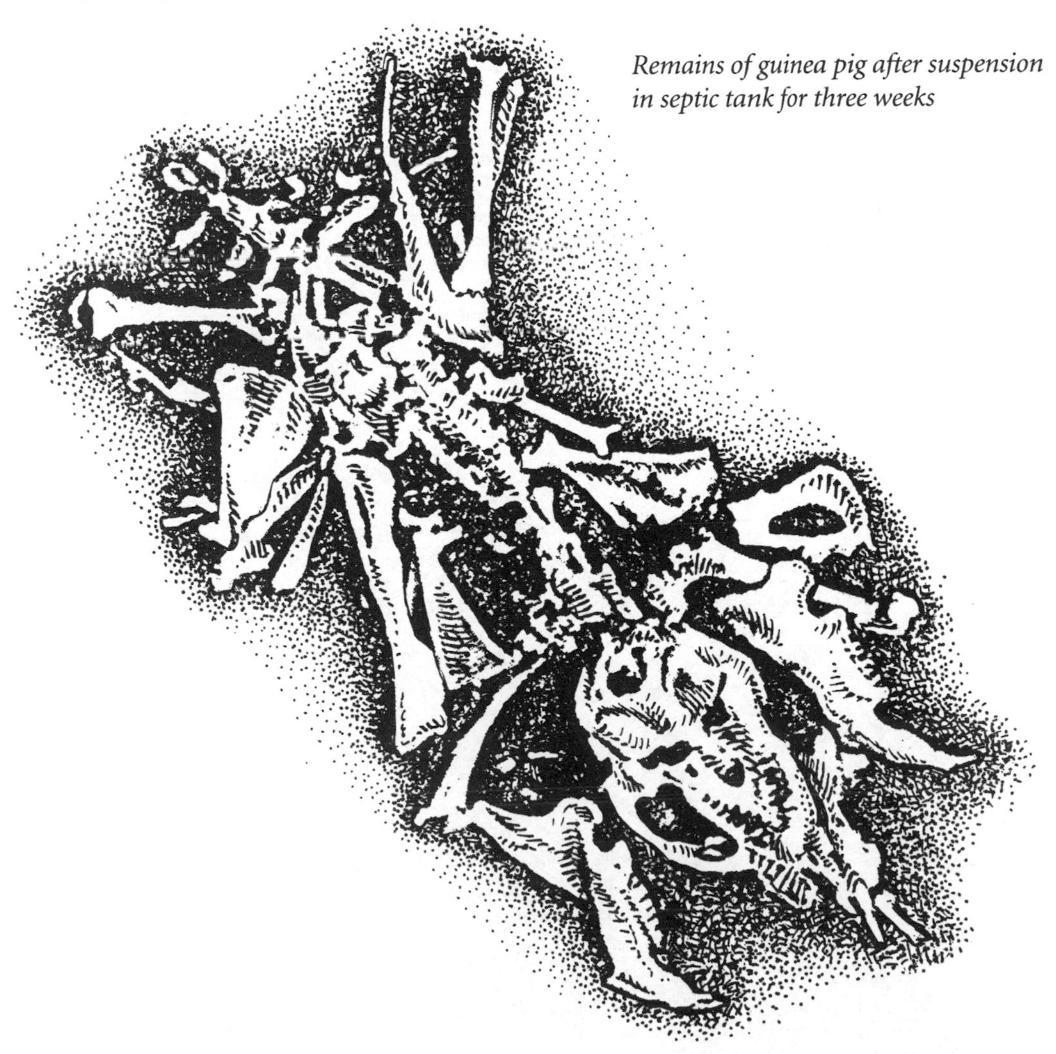

Remains of guinea pig after suspension in septic tank for three weeks

2
The Drainfield

CHAPTER TWO

The Drainfield

WHERE DOES THE WASTEWATER GO AFTER IT LEAVES THE TANK?

Most commonly, wastewater goes to a drainfield, (also called *leachfield* or *disposal field*). It can also go to a *mound (see p. 89)*, a *seepage bed (see p. 24)*, or a *seepage pit (see p. 25)*. The general and more technical term for all these methods of handling the wastewater from the tank is the *soil absorption system*. We will begin here with the most common type, the drainfield or leachfield.

WHAT THE DRAINFIELD DOES

Once sewage undergoes primary treatment in the septic tank, the clarified effluent flows to the drainfield, where it is discharged into the soil for final treatment and disposal.

Note: A typical drainfield consists of several relatively narrow and shallow gravel-filled trenches with a perforated pipe near the top of the gravel to distribute the wastewater throughout each trench. In most cases, drainfields will perform almost indefinitely if the system is designed, used, and maintained properly.

CONSTRUCTION OF DRAINFIELD

The typical drainfield trench is a level excavation, rectangular in cross section, with a level bottom. Trenches are typically 1 to 3 feet wide, 2 to 3 feet deep, and usually no more than 100 feet long. The trench is partially filled with a bed of clean gravel (¾- to 2½-inch diameter) to within 1 or 2 feet of ground surface. A single line of perforated pipe, 3 to 4 inches in diameter, is installed level (no slope) on top of the gravel and covered with an additional 2 or 3 inches of gravel.

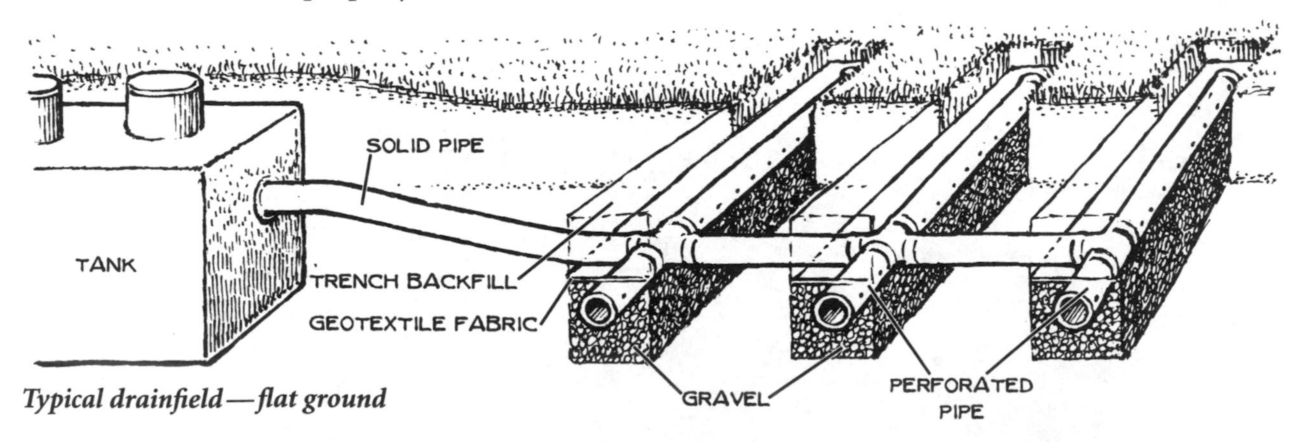

Typical drainfield—flat ground

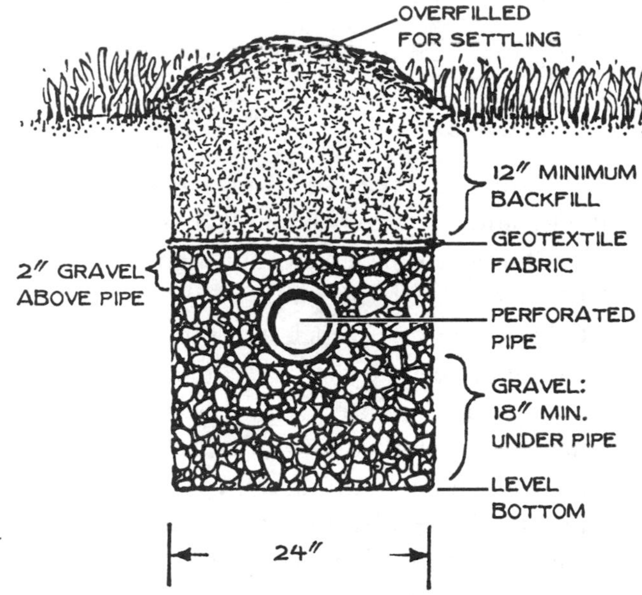

Cross section—typical drainfield. For an EPA diagram showing the pathway of air to the soil in a drainfield, see p. 159 in the Appendix.

The gravel is covered with a semipermeable geotextile fabric and the remaining 1 to 2 feet of the trench is backfilled with soil up to the ground surface. The fabric barrier keeps the backfill soil out of the gravel and prevents the fine soil particles from clogging the pores of the gravel. (Before the advent of geotextile fabric, a layer of straw or hay or untreated building paper was used for the soil-gravel barrier.)

A NOTE ABOUT DRAINFIELD DESIGNS

The size, design, and location of a drainfield for a given home depend on a variety of factors, such as local soil characteristics, the amount of wastewater flow, ground slope, and depth to groundwater or bedrock. The objective is to distribute the effluent into an area with an adequate depth of suitably permeable, unsaturated soil. The drainfield also should be located as far as possible from drinking-water wells, streams, lakes, steep hillsides, road-cuts, property lines, etc.

Specific design criteria are provided by local building or sewer codes and often vary from one locality to the next. Local health departments often require a permit before allowing any work on a drainfield and frequently can offer helpful advice. A comprehensive presentation of the many and various local design rules is beyond the scope of this book. Rather, our intent is to provide a basic understanding of what the drainfield is and how it works.

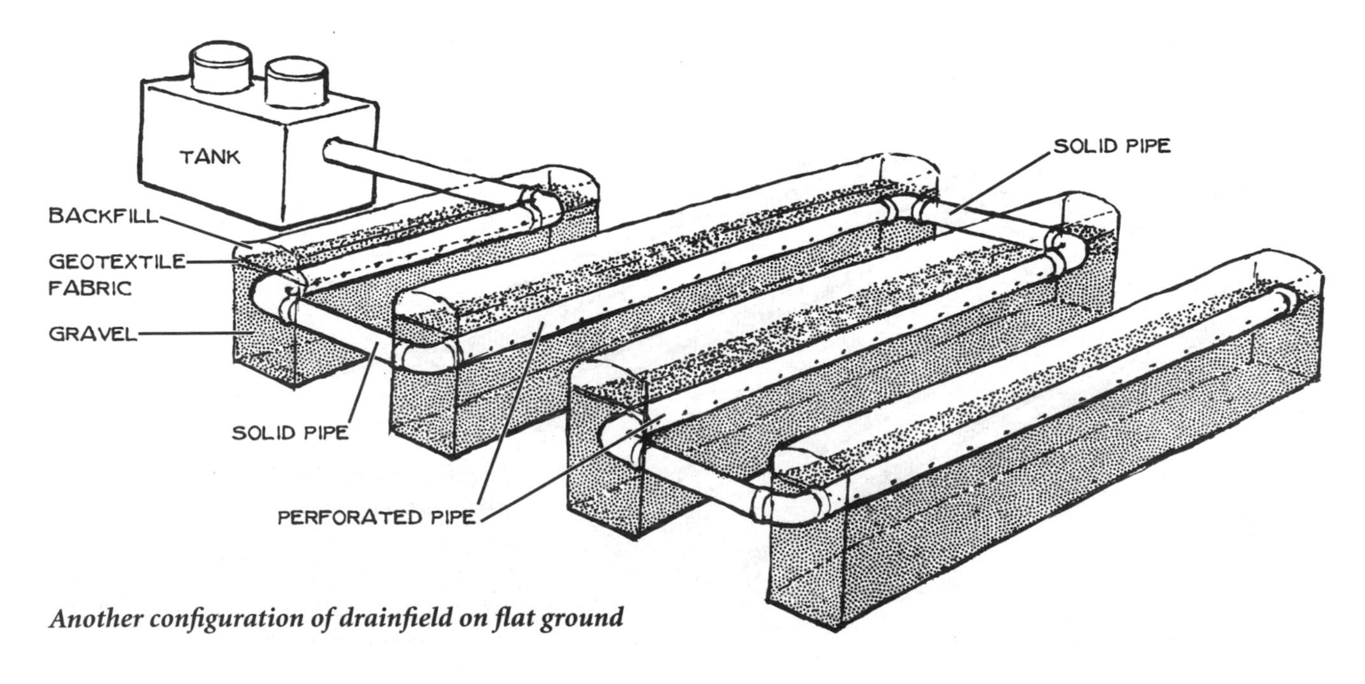

Another configuration of drainfield on flat ground

DRAINFIELD DISTRIBUTION SYSTEMS

Effluent flows from the septic tank into the drainfield through a system of watertight pipes which ultimately lead to the perforated pipes within the trenches. Various pipe connections may be included, such as tees, wyes, elbows, distribution boxes, or drop boxes. This is the drainfield distribution system.

There are two types of distribution systems: *parallel* and *serial*. With parallel distribution, effluent flows to all trenches at roughly the same time. With serial distribution, effluent flows initially to the first trench, then to the second, and so on, to each trench in sequence. The disadvantage of the latter system is that the first trench is overworked and often remains saturated, while the remaining trenches receive correspondingly less effluent. This can lead to sequential failure of the trenches. One engineer refers to serial distribution as "designed system failure."

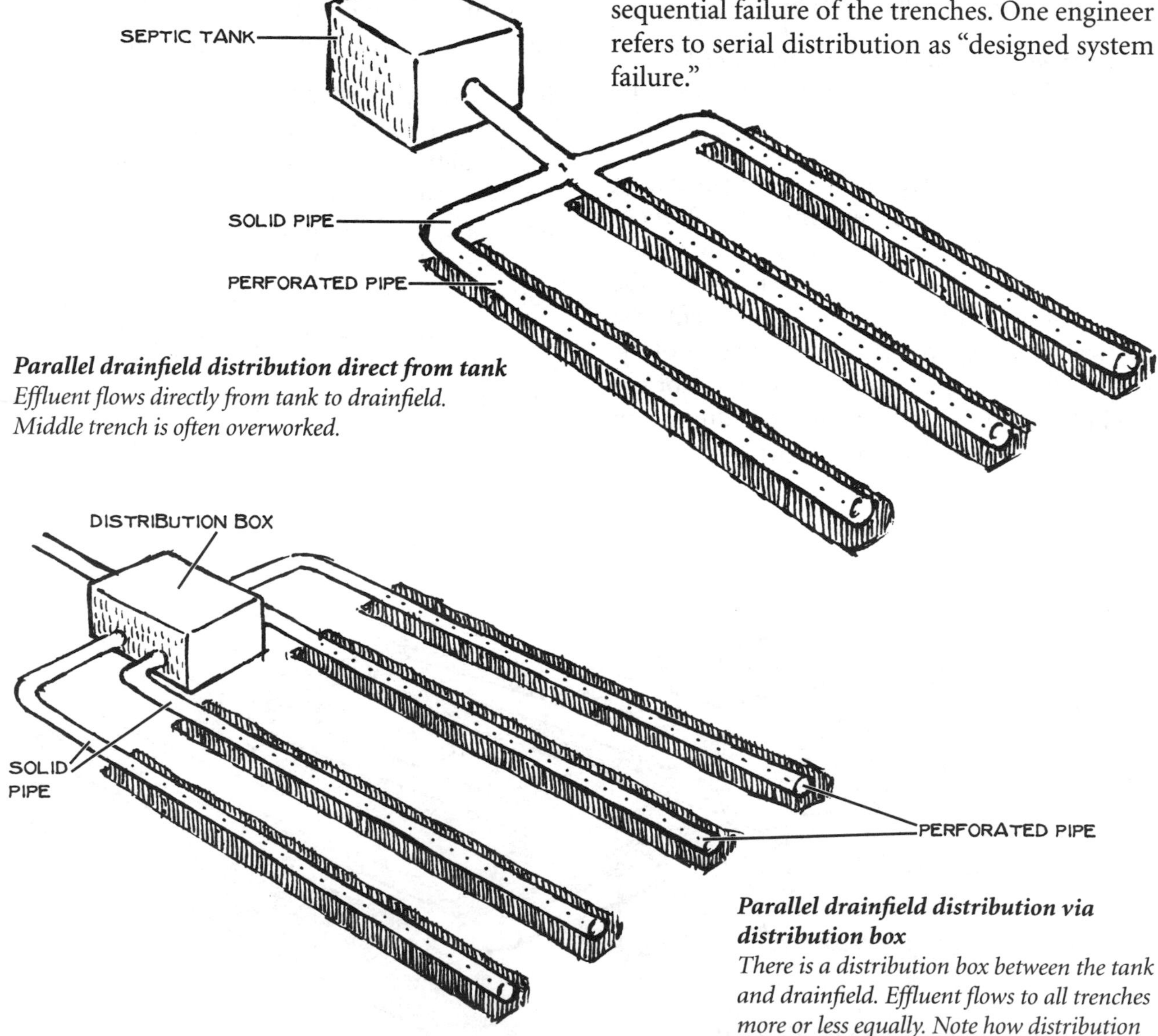

Parallel drainfield distribution direct from tank
Effluent flows directly from tank to drainfield. Middle trench is often overworked.

Parallel drainfield distribution via distribution box
There is a distribution box between the tank and drainfield. Effluent flows to all trenches more or less equally. Note how distribution here differs from that above.

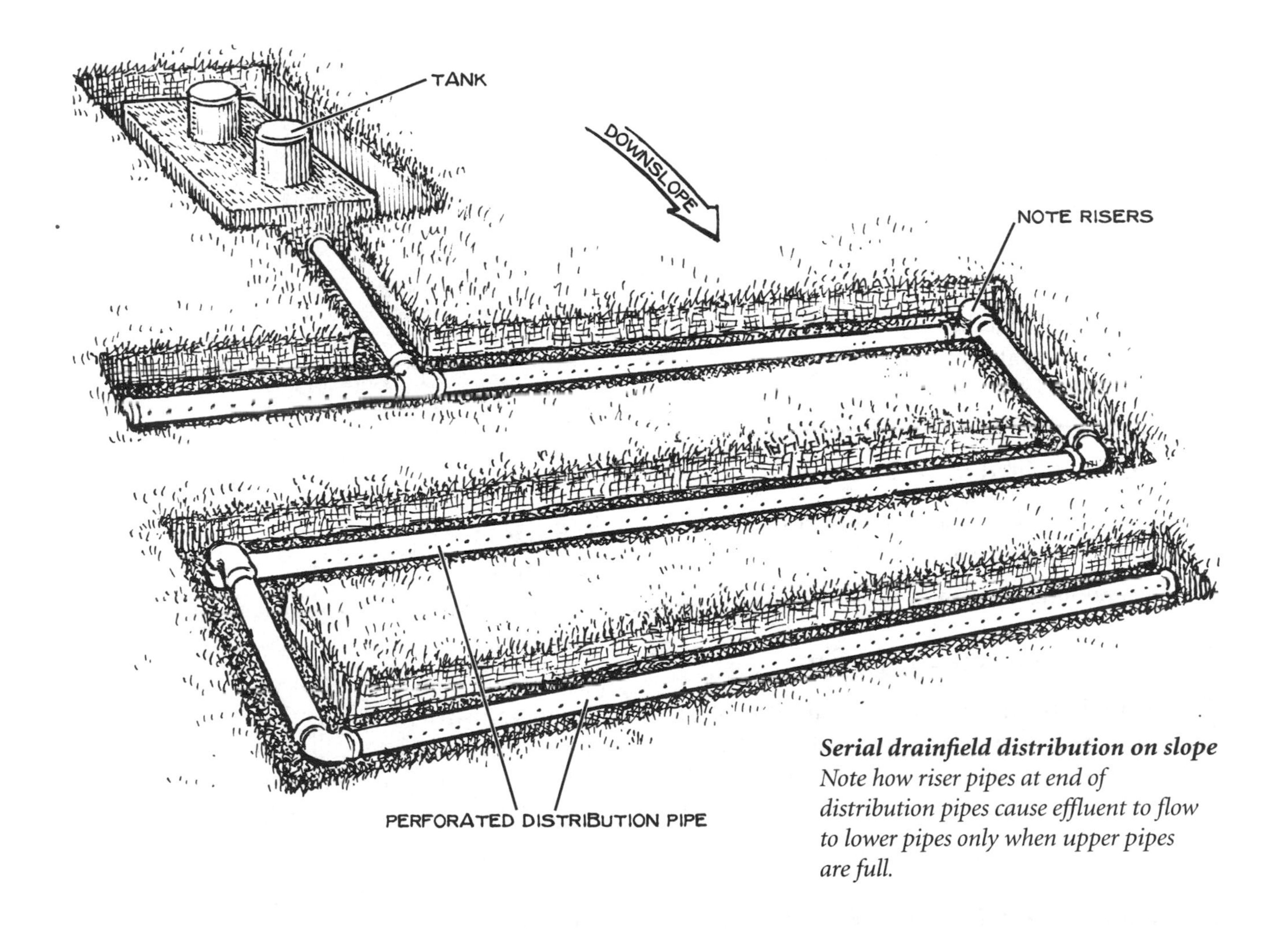

Serial drainfield distribution on slope
Note how riser pipes at end of distribution pipes cause effluent to flow to lower pipes only when upper pipes are full.

HOW THE DRAINFIELD WORKS

The drainfield provides both *disposal* and *treatment* of the septic tank effluent. Effluent flows from the septic tank to the drainfield through a watertight pipe and is then distributed within the drainfield trenches through perforated pipes in the gravel. The effluent flows through the gravel filling and then seeps (infiltrates) into the soil beneath and beside the trench. Here the main purification of the wastewater takes place through filtration and biological activity as it infiltrates through the *biological mat (see p. 19)* on the sidewalls and at the bottom of the trench and then percolates through unsaturated soil. (However, in seasoned fields, there is practically no percolation through the trench bottom.) The purified liquid then eventually evaporates, is taken up by plants, or percolates into the groundwater.

MOVEMENT OF EFFLUENT

Once the effluent leaves the septic tank, it flows:

- to the trenches through a watertight distribution system
- into the trench gravel through perforated pipes
- into the soil through the infiltrative surfaces (gravel-soil interface)
- through unsaturated (or saturated) soil beside and beneath the trench

Finally, it is either taken up by plants or percolates to groundwater.

TREATMENT OF EFFLUENT

Effluent gets treated in the drainfield:

- as it infiltrates into the soil: the biomat is the tool. *(See adjacent column.)*
- as it percolates through the soil

FUNCTION OF THE GRAVEL

The gravel in the drainfield maintains the structure of the trenches and helps distribute the effluent to the infiltrative soil surfaces. Also, the porosity of the gravel provides temporary storage capacity during peak flows.

SECOND THOUGHTS ON GRAVEL

Engineer Randy May believes there are a number of things wrong with the use of gravel in drainfields. Although the gravel maintains the structure of the trenches, it also is filling ⅔ of the available storage capacity in the drainfield with rock. The gravel itself provides little treatment of effluent and, moreover, may cause compaction in the trench bottom due to its weight and fines (if present) in the gravel. *(See p. 22 regarding shallow drainfields, which use no gravel, and p. 23 regarding Infiltrator™ Chamber Leaching Systems.)*

GRAVITY RULES!

THERE ARE TWO BRILLIANT THINGS ABOUT A PROPERLY FUNCTIONING GRAVITY-FED SEPTIC SYSTEM:

- THE POWER IS GRAVITY, THE EARTH'S MAGNETIC FORCE — PULLING WATER AND WASTES DOWNWARD. NO PUMPS, NO ELECTRICITY.
- THE CLEANSING AGENT IS THE EARTH, WHERE MICROORGANISMS, NATURALLY PRESENT IN THE SOIL FILTER, FEED ON AND PURIFY SEPTIC TANK EFFLUENT, INCLUDING DISEASE-CAUSING ORGANISMS.

IT IS AN ELEGANT DESIGN, WORKING SILENTLY UNDERGROUND, AND REQUIRES ONLY A MINIMUM OF MAINTENANCE TO KEEP IT FUNCTIONING. IT IS PRACTICAL, FUNCTIONAL, AND ECOLOGICAL. IT'S WORTH KNOWING HOW TO BE AN INTELLIGENT STEWARD OF THIS LIVING SYSTEM.

THE BIOMAT

What It Is

The *biomat* (biological mat) is a black, jelly-like mat that forms along the bottom and sidewalls of the drainfield trench. It is composed of anaerobic microorganisms (and their by-products) that anchor themselves to soil and rock particles. Their food is the organic matter in the septic tank effluent. Since the biomat has a low permeability, it slows down the rate of flow out of the trench into the drainfield soil and also serves as a filter to provide effluent treatment. With a well-developed biomat, wastewater may be temporarily ponded in the drainfield trench, yet the soil a few inches outside the trench will be unsaturated.

Biomat Formation

The biomat forms first along the trench bottom and then up along trench walls. It has less permeability than fresh soil, so incoming effluent will pond over the biomat and trickle along the trench bottom to an area where there is little or no biomat; eventually the biomat will line the bottom of the trench and form up along the walls as well.

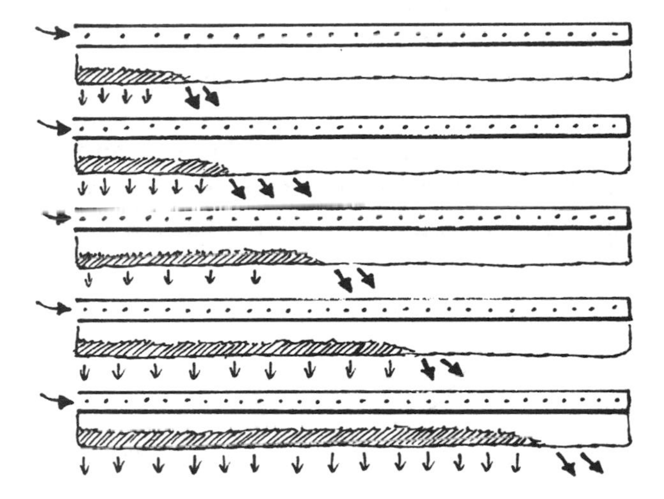

Progressive development of biomat on infiltrative surface of drainfield

The Biomat Is a Living Filter

Through filtration and biological activity, the biomat is very effective at removing viruses and, in fact, filters out pathogenic bacteria and parasites.

Biomat in Equilibrium

In the trench, in a saturated environment, the anaerobic organisms in the biomat feed on the organic material in the wastewater; this causes the biomat to grow thicker and decreases permeability. *On the soil side* of the biomat, in an unsaturated environment, aerobic soil bacteria feed on and break down the biomat. In ideal conditions, these two processes go on at the same time, so the thickness and permeability of the biomat stay fairly consistent. Beautiful!

Biomat Problems

Since the biomat is alive, its equilibrium can be upset. Failure to regularly pump out the septic tank can result in an excess of organic material (food) to the biomat organisms, causing excessive growth and, therefore, reduced permeability. In saturated soils, aerobic conditions no longer exist, and controlled breakdown of the biomat by aerobic soil bacteria will not occur.

If the septic system is poorly maintained, too much wastewater is flushed down the drain, or the drainfield soil remains saturated, the biomat eventually will grow too thick and dense, and the effluent sent to the drainfield will exceed the amount that can filter through the biomat. This can result in excessive ponding in the trenches, backflow into the septic tank (and possibly also into the house), or surfacing of effluent above ground over the drainfield—in other words "failure."

See pp. 30–35 for details on how soil cleans water.

MAINTAINING A STABLE BIOMAT

Recent studies have shown that it is important to distribute the waste as evenly as possible along the entire infiltration zone. In many older systems, the pipe size is too large and the waste only flows a short way along the drainfield trench, dumping most of the effluent at the beginning of the trench, and none further along. When this occurs, the beginning of the trench eventually forms too thick a biomat of bacteria and other microorganisms, and plugs up. The flow then proceeds further down the trench until that segment plugs up. When the whole system becomes plugged, the drainfield fails. Dosed-flow drainfields *(see pp. 87–89)* can solve this problem. However, even in gravity-flow drainfields, with the right type soil, piping, and media for good distribution (as well as periodic pumping of solids from the tank), the biomat builds up to a stable layer, and the drainfield can go on treating the waste almost indefinitely.

DUAL DRAINFIELDS

Dual drainfields are considered alternative systems in some areas, yet in other areas they are standard practice and are either required or at least recommended for all systems. A dual drainfield does not solve the problem of inadequate site conditions. Site and soil conditions suitable for a conventional drainfield are needed. However, a dual drainfield does provide improved treatment and disposal performance over the standard drainfield system.

What Is a Dual Drainfield?

The system consists of two independent drainfields with each one designed to accommodate the entire wastewater flow. They may be built with intermeshing parallel trenches or side by side. There is a diversion valve on the line from the tank, so effluent flow can be directed to one field or the other at any one time.

The Diversion Valve

The valve is periodically rotated, typically on a yearly basis, bringing the second field into use and allowing the first field to rest. This resting period allows the microorganisms in the drainfield soil to break down organic matter (the biomat) that has accumulated in the drainfield trench/soil interface during the use period, thus partially restoring the infiltrative capacity of the drainfield.

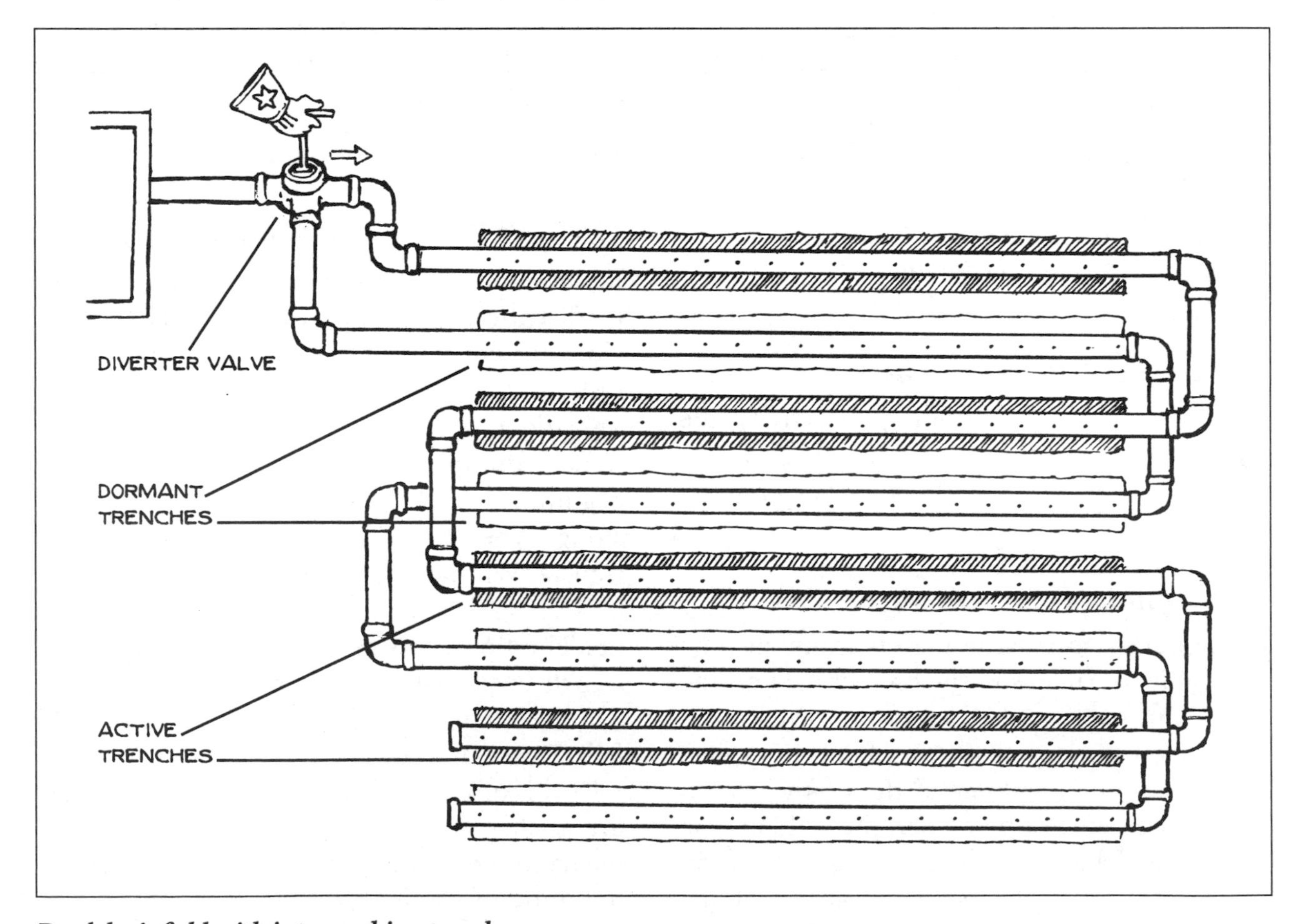

Dual drainfield with intermeshing trenches
Flow to two different sets of trenches is controlled by diversion valve. (See next page.)
Note here how flow goes to every other trench (shaded areas).

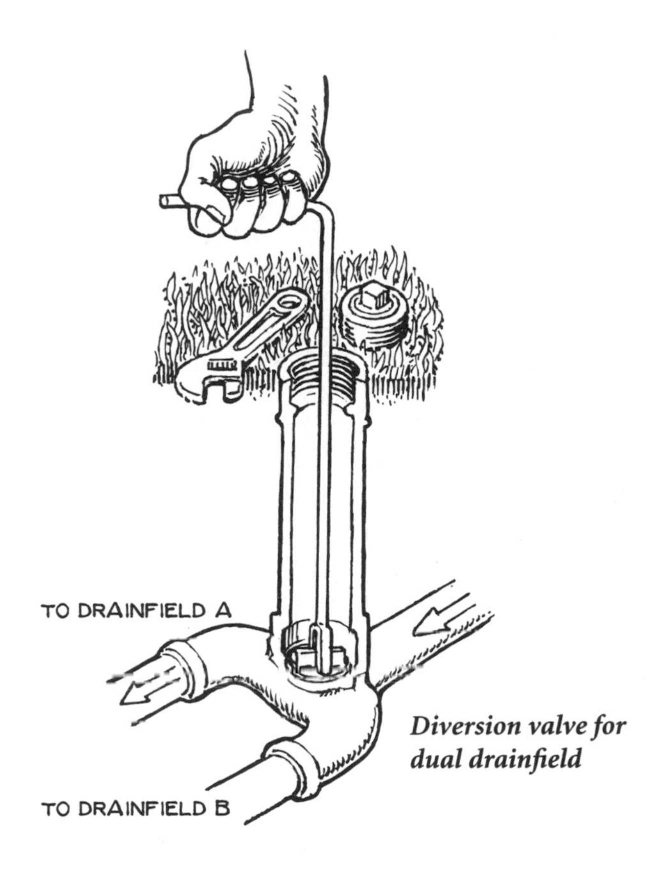

Diversion valve for dual drainfield

Homeowners Take Note!

The weak point with this system is the inability of the average homeowner to remember to turn the valve at the appropriate time. Write it on your calendar. One homeowner told us she does it twice yearly, on the 4th of July and New Year's Day.

Homeowners' Tip

With intermeshed drainfields, as above, you should switch drainfields at intervals so that the resting field is always moist enough to keep the bacteria in the biomat alive. *(See p. 18.)* If the dual fields are built as two entirely separate fields, dose the resting field periodically to keep soil microorganisms alive.

Dual Drainfields

Local regulations often require, or at least recommend, dedication of a "replacement" or "reserve" area on the property. Then, if the original drainfield fails, there is space available for its replacement. A dual drainfield system essentially calls for construction of the "replacement" and primary drainfield at the same time.

The principle involved here—two independent drainfields utilized on an alternating basis—can be applied to all types of drainfield systems, both conventional and alternative.

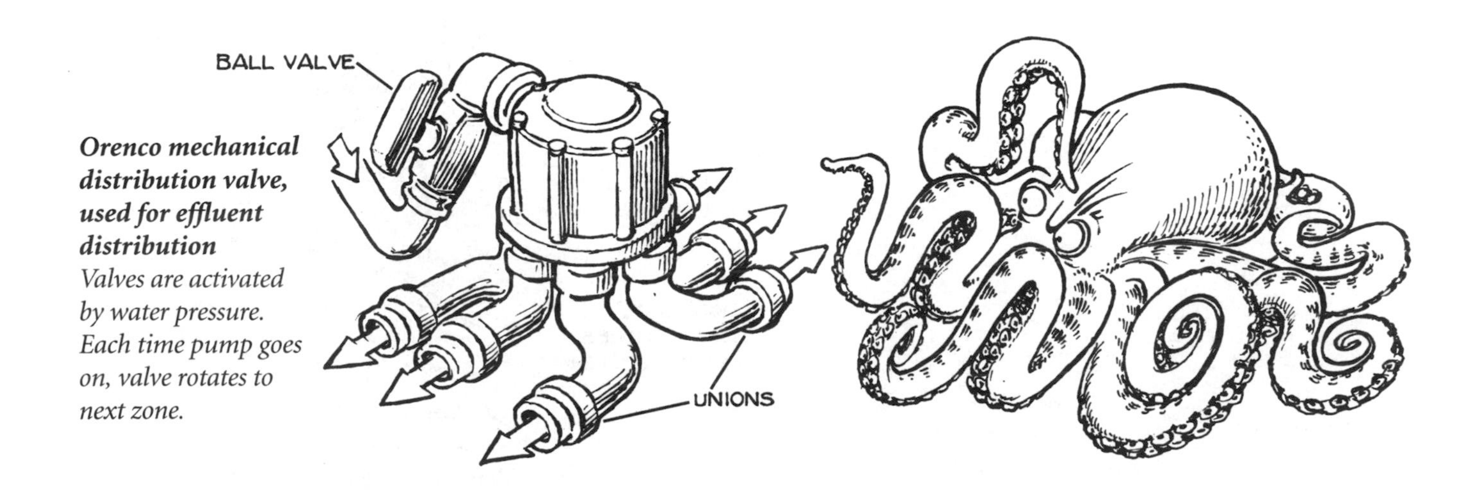

Orenco mechanical distribution valve, used for effluent distribution
Valves are activated by water pressure. Each time pump goes on, valve rotates to next zone.

THE NEW (& OLD) SHALLOW DRAINFIELD

There are millions of microorganisms in the trench of a typical drainfield. But there are billions of these organisms in the top foot or so of soil (98% of soil life is in the top 16 inches). Shallow, gravel-less drainfields, which utilize the cleansing properties of the top few inches of soil (as well as plant roots) in removing nitrogen and phosphorus, are now in use throughout North America. The EPA estimates that 750,000 shallow drainfields have been installed in the United States in the past 15 years.

Historical Perspective

In the 1890s, when drainfield trenches were dug by hand, shallow trenches were standard. Many drainfields used during this time were overplanted with a garden — thus assuring the plants a constant supply of nutrients and water. When systems were loaded with more effluent than could be handled, ponding occurred above ground.

SHALLOW DRAINFIELDS, WITH PURIFICATION BY SOIL LIFE, PROVIDE *TREATMENT*, AS OPPOSED TO DEEPER DRAINFIELDS, WHICH PROVIDE *DISPOSAL* OF EFFLUENT.

With the advent of the backhoe, and its deep-digging capacity, trenches became deeper. "Out of sight, out of mind." Less ponding, but also less cleansing action from the top six inches of soil. Now things seem to have come full circle, but with the technical advantage of leaching chambers and pressure dosing. Also of interest: the standard distance required between typical gravel-filled trenches is the distance necessary for a backhoe to operate.

The Infiltrator® Chamber

The Infiltrator® Chamber System is installed in a shallow drainfield that replaces the typical gravel and pipe drainfield. It consists of a series of interlocked high-density polyethylene chambers that rest on the bottom of a shallow trench. The chambers have louvered sidewalls and are open at the bottom, allowing the effluent to

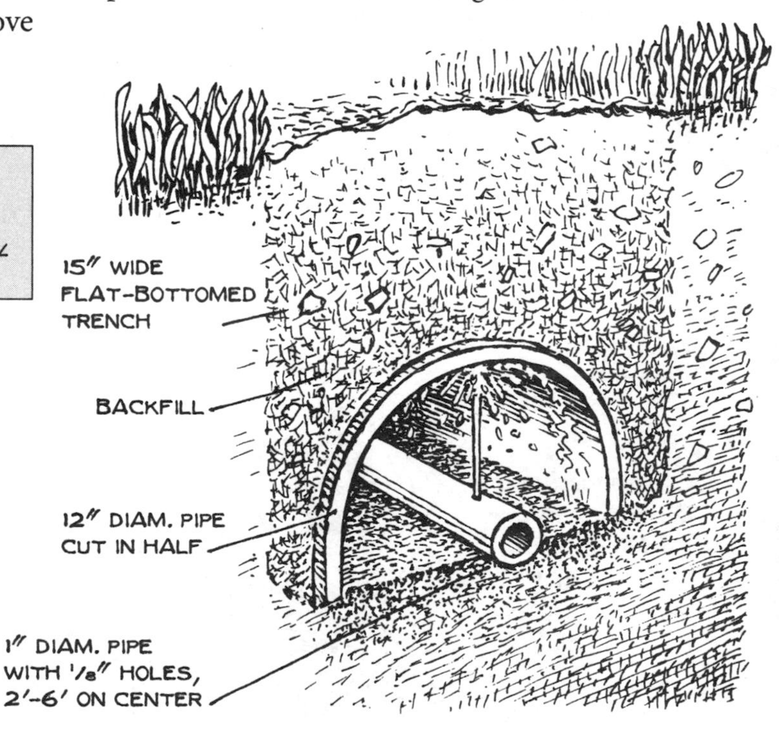

Shallow drainfield with pressure dosing
Large piece of pipe cut in half provides chamber. Bottom of trench is 10″–12″ deep.

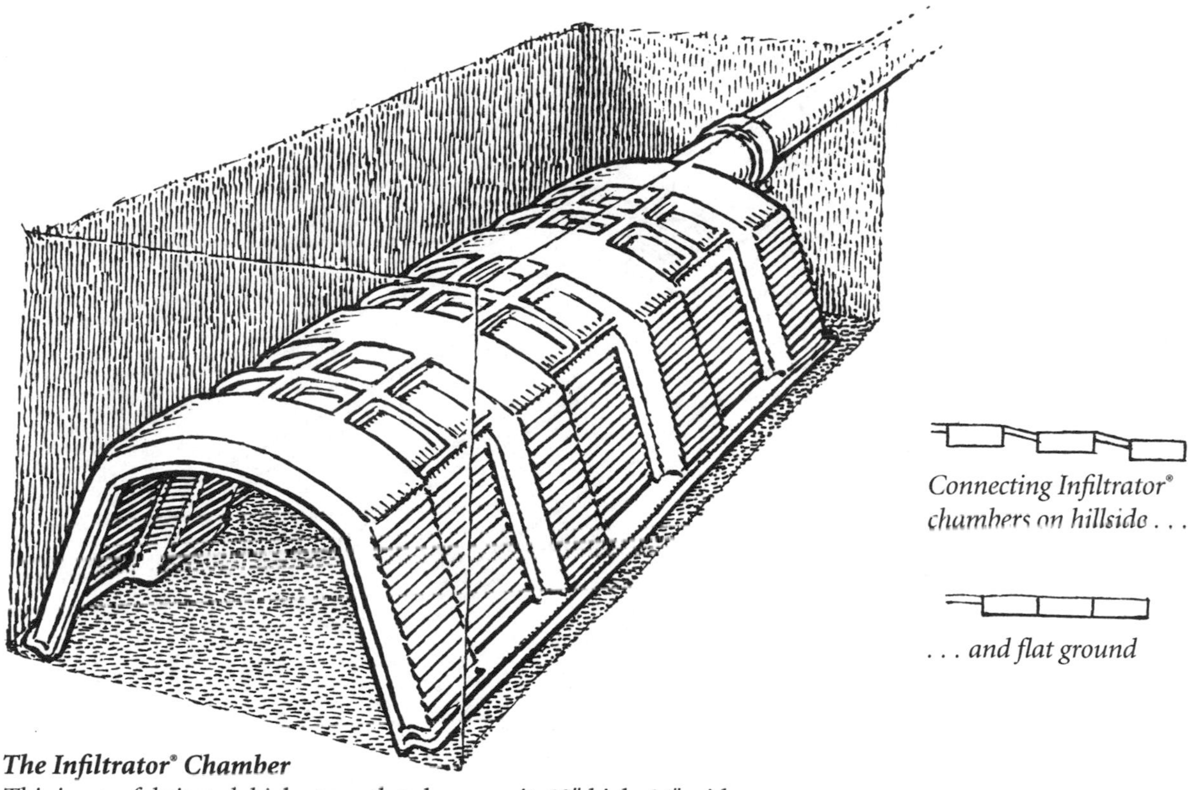

Connecting Infiltrator® chambers on hillside . . .

. . . and flat ground

The Infiltrator® Chamber
This is a prefabricated, high-strength polymer unit, 12" high, 34" wide, 6.25' in length. Sections are locked together and placed on bottom of drainfield trench, then backfilled with soil. No gravel is used.

pass directly into the soil. There is no gravel or horizontal pipe. The large volume of the chamber not only provides additional storage capacity for peak flows of incoming waste, but provides a ready source of oxygen above the vadose zone (permeable soil that air can pass through *[see Appendix, p. 159]*). Many states and counties allow a drainfield of only half the size when using leaching chambers. The chambers can be installed in shallow trenches with as little as 6" of soil over the top of the chamber; this allows air to reach the very area of the soil with the healthiest microbial population, thus insuring the best treatment of waste.

Note: Pressure dosing is often used with shallow trenches. This involves a pump and pump chamber *(see p. 88)* and ensures that the effluent is distributed evenly throughout the drainfield, thus minimizing clogging. Also, see p. 88–89 on pressure dosing with drip irrigation.

Shallow drainfields are usually installed by hand (a shovel) and do not require heavy equipment such as gravel trucks on site, thus avoiding compaction of the soil in the drainfield. Shallow drainfields are often owner-installed when a conventional drainfield has failed, and/or over-restrictive regulations or over-designing engineers are present. The unit shown above is manufactured by Infiltrator Systems, Inc., Old Saybrook, Connecticut. (*Note:* Infiltrator now has a new smaller chamber called the Quick4 Equalizer 24 Chamber.)

www.infiltratorsystems.com / 800-221-4436

A SHALLOW DRAINFIELD CAN REDUCE NITROGEN BY HALF.

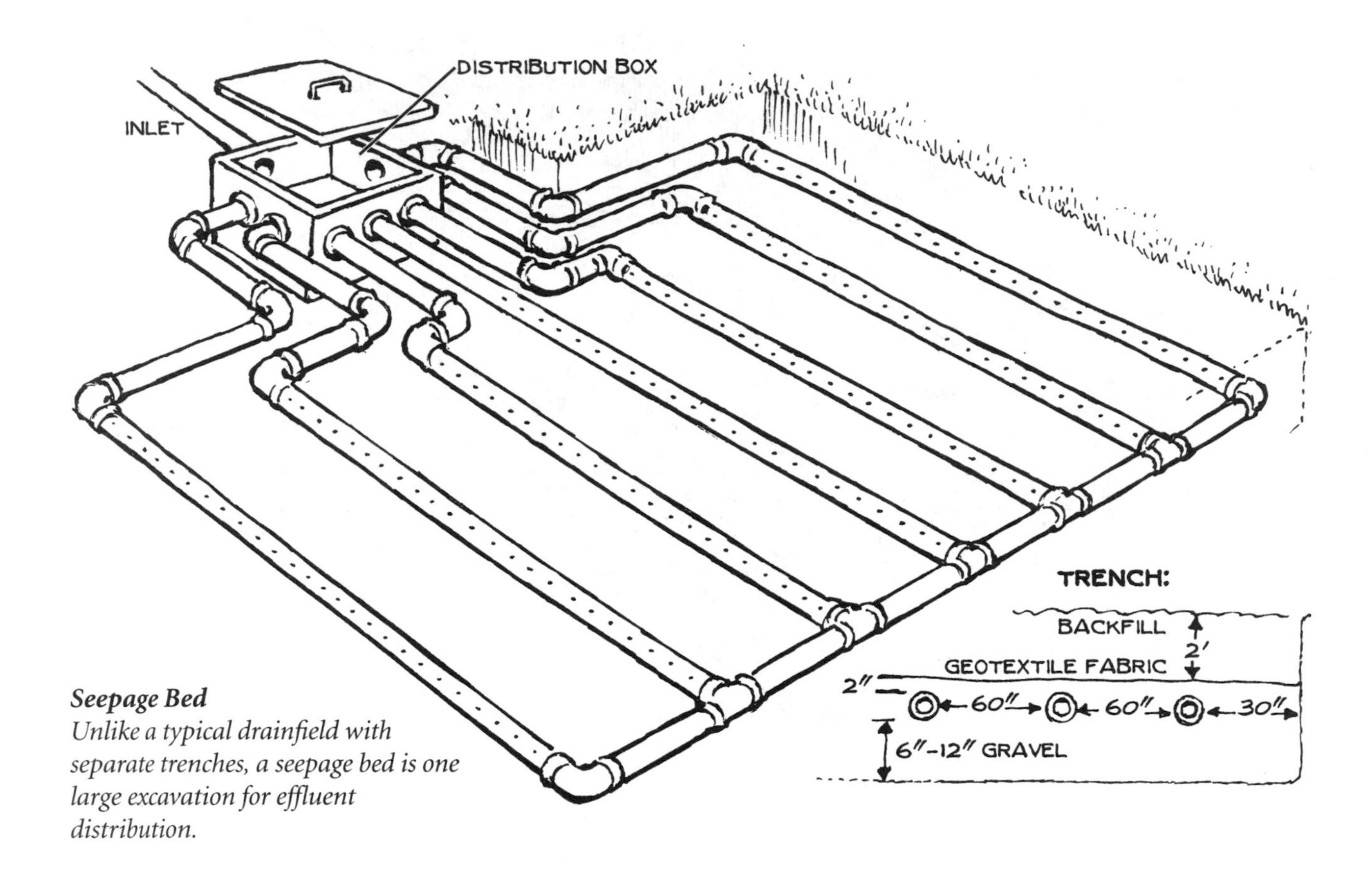

Seepage Bed
Unlike a typical drainfield with separate trenches, a seepage bed is one large excavation for effluent distribution.

Other Absorption Systems

Seepage beds and *seepage pits* are soil absorption systems and are used instead of typical drainfields. A *cesspool* is a tank with no absorption system. Each of these three has its own drawbacks compared to the systems so far described. However, since many of these have been built and are still in use, we will describe each of the three.

SEEPAGE BEDS

Generally, any drainfield trench wider than 3 feet is considered a seepage bed. Unlike a trench-type drainfield, a seepage bed drainfield typically consists of just one large excavation. The bottom is covered with a layer of gravel, and multiple distribution pipes are laid out on this gravel bed, with spacing of 3 to 5 feet between pipes. As in an individual drainfield trench, the pipes are covered with some gravel and a protective barrier of geotextile fabric (or straw and untreated building paper) and a backfilling of soil up to ground surface.

The importance of flat terrain: Seepage beds of a gravity-based system should be as level as possible. If pressure dosed, where effluent goes out in periodic surges, level is less important.

What's Good About a Seepage Bed

- It typically requires less land area than a comparable trench system.
- It is less expensive.

What's Bad About a Seepage Bed

- In comparison to a trench system of similar size, it has considerably less infiltrative surface area (the gravel-soil interface where disposal into the soil and treatment through the biomat occurs). Conventional drainfields have infiltrative surface area on all trench sidewalls, whereas the bed has comparatively little sidewall area. The sidewalls are particularly important for long-term performance, as the bottom surface areas are the first to become clogged due to biomat growth.
- Use of a bulldozer or bucket loader in construction compacts the soil on the bottom of the bed, destroying soil structure.

SEEPAGE PITS

A seepage pit is, in essence, a vertical leachline, consisting of a deep hole with a porous-walled inner chamber and a filling of gravel between the chamber and the surrounding soil. Seepage pits are typically 4 to 12 feet in diameter and 10 to 40 feet deep. Septic tank effluent enters the inner chamber and is temporarily stored there until it gradually seeps out and infiltrates into the surrounding sidewall soil.

Note: Overall, seepage pits are generally considered undesirable in that they deliver effluent directly to groundwater with no treatment. The advantages apply only in arid regions, where groundwater is very low.

The wall, along with the surrounding gravel filling, provides structural support for the pit. It is 1 to 2 feet smaller in diameter than the pit, and may be constructed of pre-cast concrete rings or a cylindrical wall of brick or concrete block. For outward seepage of the effluent, concrete rings have pre-cut holes or notches; bricks or blocks are laid up with staggered joints. In a smaller diameter pit, the inner chamber may consist of a large-diameter perforated pipe standing on end. The bottom of the pit is usually covered with 6 to 12 inches of gravel.

The top of the chamber should have a cover of suitable strength, such as reinforced concrete or thick steel plate, extending about 12 inches beyond the edge of the pit. The cover should be equipped with a manhole for inspection purposes.

As with drainfield trenches, seepage pits experience progressive biomat growth. As the biomat grows denser in the lower level, the effluent rises to a higher level, where it filters through the as-yet-unclogged sections of the sidewall.

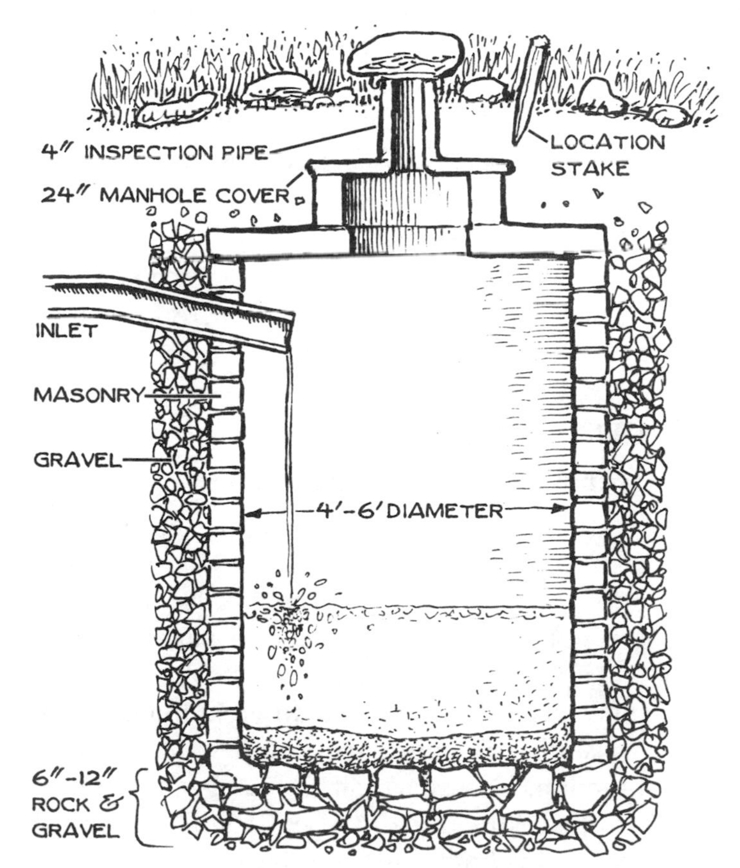

Seepage pit
A deep chamber is excavated for effluent disposal. Effluent seeps through porous masonry into surrounding gravel.

What's Good About Seepage Pits

Seepage pits have certain advantages over drainfield trenches:

- They take up less surface area on a lot.
- They are not as seriously affected by rain and are not as subject to invasion by tree roots.
- The level of liquid fluctuates up and down significantly more than in a drainfield trench. This provides aeration of the wetted sidewall soil, allowing it to remain more permeable.

What's Bad About Seepage Pits?

- Seepage pits are suitable for use only where there is very deep soil of good permeability and considerable depth to groundwater. (Health codes generally require that seasonal high groundwater level be at least 3 to 4 feet below the bottom of the pit.) Such conditions are found in the arid regions of the southwestern United States but are unusual elsewhere.
- Because of their construction, they do not utilize the cleansing action of microorganisms in the top few inches of soil.

CESSPOOLS

Seepage Pits or Cesspools?

In the old days, before septic tanks were common, sewage from homes was often disposed of in covered vertical pits called cesspools. It is not uncommon to hear a seepage pit called a cesspool, or vice-versa. Nevertheless, seepage pits and cesspools are two different things. They are similar in construction but serve different functions. A seepage pit is used for disposal of clarified effluent from a septic tank, whereas a cesspool is both tank and drainfield (the latter being the soil around the cesspool).

The Earliest "Septic System"

The cesspool was an early predecessor of the current-day septic system, providing a subsurface system for disposal of waterborne sewage. Use of cesspools was primarily driven by convenience ("out of sight, out of mind") and their location governed mainly by the nearest available land. In fact, cesspools were even constructed in the basements of urban buildings. This is in obvious contrast to current septic system designs, based on soil and groundwater conditions, and intended for long-term use.

Today, cesspools are viewed as undesirable by public health officials, due, in part, to historical experiences with overused and failed systems, as well as increased concern for groundwater protection. In most areas, cesspools are prohibited by local health and building codes. Consequently, new construction of cesspools is relatively uncommon. Nevertheless, numerous cesspools are still in active use, most commonly in rural areas, but also in some older homes in suburban communities.

What Is a Cesspool?

A typical cesspool is a cylindrical hole in deep soil, several feet in diameter. In many, there is a porous inner wall of stone, masonry, precast concrete rings, or other material strong enough to shore up the soil. The outer surface (between the masonry wall and the outer soil wall) is filled with gravel. There is a concrete lid and, on top of that, soil backfilled to grade.

Raw wastewater flows into the top of the inner chamber. The inner chamber retains (and partially digests) the solids, and the effluent seeps through to the gravel-filled outer chamber, and then into the surrounding soil.

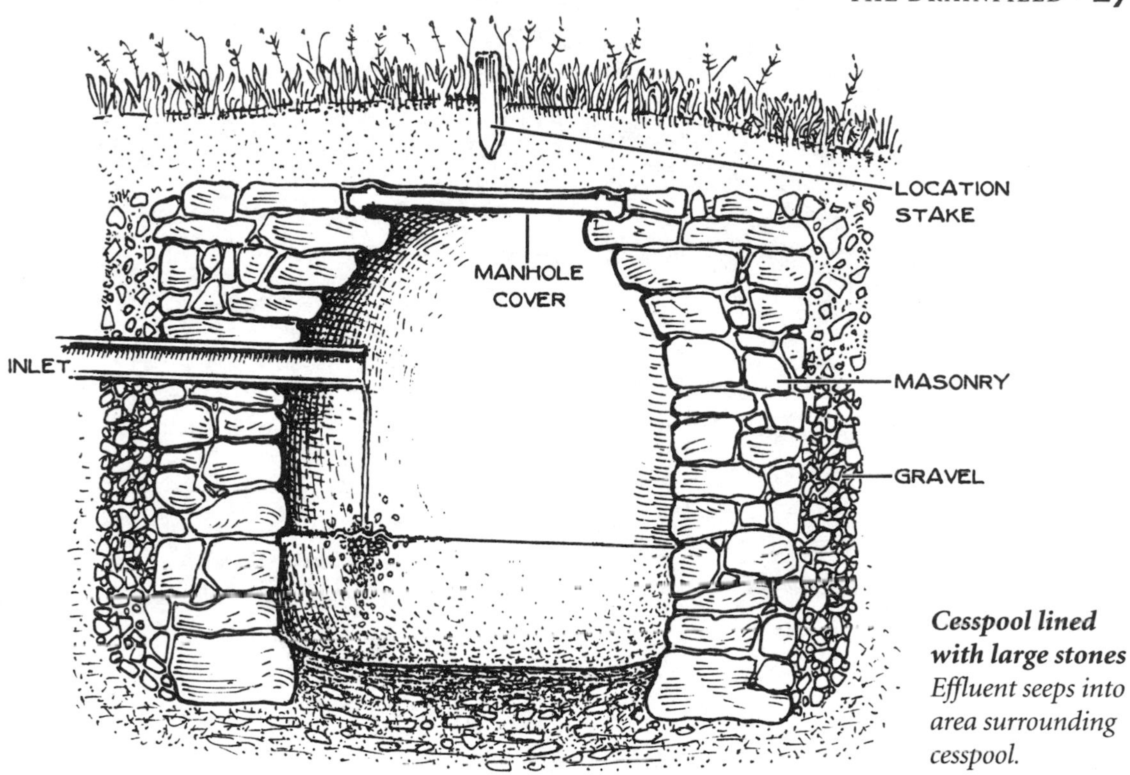

Cesspool lined with large stones
Effluent seeps into area surrounding cesspool.

"Functionally, the inner wall of the cesspool is a leaky septic tank, and the outer soil wall is the subsurface disposal field."

–*Septic Tank Systems: A Consultant's Toolkit,* John H. Timothy Winneberger, Butterworth Publishers, Boston, 1984

Requirements for Cesspools

Design of cesspools depends upon the ability of the soil to absorb water. They should not be used in porous soil or where groundwater may come to within 3 to 4 feet of the bottom. They should also be downhill and 200 to 500 feet away from wells or springs used for drinking water.

> **FULL DISCLOSURE IN HOME SALES**
>
> IN RECENT YEARS, DISCLOSURE OF THE CONDITION OF THE SEPTIC SYSTEM OF A HOME BEING SOLD HAS BECOME A LEGAL ISSUE. THE ARGUMENT THAT THE SELLER IS OBLIGED TO INFORM THE BUYER OF ANY SEPTIC SYSTEM MALFUNCTION HAS BEEN MADE SUCCESSFULLY IN COURT. THUS, IT BEHOOVES THE SELLER TO MAKE FULL DISCLOSURE TO THE BUYER AS TO SEPTIC SYSTEM CONDITION. (IT IS ALSO THE FAIR THING TO DO.) LIKEWISE, IT BEHOOVES THE BUYER TO REQUEST DISCLOSURE FROM THE SELLER. THIS IS ESPECIALLY IMPORTANT IF THE BUYER HAS NO PRIOR KNOWLEDGE OF SEPTIC SYSTEMS.

SUMMARY OF SEPTIC SYSTEM FUNCTIONS

A septic system is a sewage disposal system. Waterborne wastes leave the house via the drains and enter the septic tank, where primary treatment occurs. This is under anaerobic (without air) conditions. Here the solids are separated from the liquids and scum and sludge layers accumulate. These must be removed periodically. *(See pp. 49–52.)* The liquids (effluent) flow from the septic tank into the (underground) drainfield, where they are brought into contact with the soil community. The soil provides secondary treatment, further cleaning the sewage. The water and nutrients are thus brought back into the earth, helping to maintain a healthy soil structure for plant growth, and completing the cycle.

3
4

3
The Soil

CHAPTER THREE

The Soil

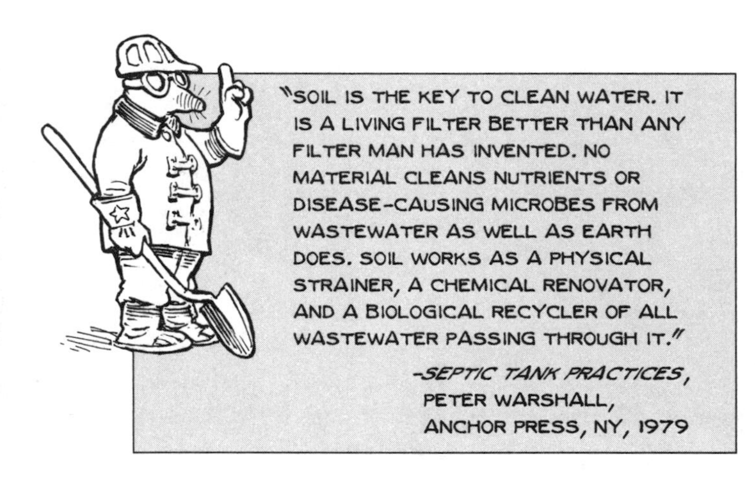

THE SECOND NATURAL WONDER OF SEPTIC SYSTEMS

The first wonder of the conventional septic system is the mechanism of gravity power. *(See p. 18.)* No hardware, no motors. The second wonder is the amazing function of naturally occurring soil microorganisms in purifying septic effluent. Underground, a host of microscopic life forms are busy filtering, feeding on, and thereby purifying septic tank effluent.

SOIL BASICS

We will define some basic terms and attempt to give you a sense of these unseen, largely unknown, natural subterranean life forms that are working night and day on your behalf. Although this information may not seem as practical as other material in the book, it may have its useful aspects: if you understand the soil's life forms and their functions, it may help you to make better-informed decisions about what goes down the drain and what you can do to maintain a healthy system. *(See Chapter 4, "Down the Drain.")* Further, understanding a bit about the cleansing action of the soil is fascinating even to non-engineers—an example of a powerful force existing in nature that humans have been able to utilize so well.

Suitable soil is an effective treatment medium for septic tank effluent. The soil community consists of billions of microorganisms—including bacteria, protozoa, fungi, molds, and other organisms—living and interacting with the minerals, organic matter, water, and air which comprise the soil. This complex biological community, as well as the physical structure of the soil itself, provides the mechanisms by which the wastewater is purified as it percolates through the soil and returns to the underlying groundwater, or is taken up by plants, or is evaporated.

SOIL TEXTURE

Soil texture is important because of the relationship between soil particle sizes and the ability of the soil to absorb and purify wastewater.

Analysis of site soil conditions is of essential importance in septic system design, but is complex and beyond the scope of this book. *(See bibliography, pp. 166–173 for references.)* However, here are a few general principles.

Coarse-Textured Soil

A coarse-textured soil, such as coarse sand, provides good *percolation* of septic tank effluent since the water can move more rapidly through the coarse soil than through a fine-textured or clay soil. However, in order to provide adequate *treatment,* a coarse soil must be deep. *Note:* many states prohibit drainfields in sandy soil and require elevated drainfields, or mounds, with imported soil. *(See pp. 89–90.)*

Fine-Textured Soil

A fine-textured soil can provide adequate treatment in the top 16 inches of soil depth, but the water generally moves slower than in coarse soil. Many clay soils swell when wet and take a long time to become semisaturated or unsaturated, and in some cases the small pore spaces may "slam shut" due to expanding clay particles. Once this happens, it is often mechanically impossible for the pore spaces to reopen.

Ideal Soil

The ideal soil for effluent absorption and treatment contains a balanced mix of coarse and fine particles, such as a loam or clay loam.

WATERBORNE DISEASES

We'll start with the worst-case scenario. If the septic system and the soil do not perform adequately, there are some major diseases that can be spread in sewage-contaminated water and soil by pathogenic viruses, bacteria, protozoa, or tapeworms.

Viruses can cause polio, hepatitis A, and epidemic and sporadic viral gastroenteritis. Viral agents can be transmitted in water or food contaminated by sewage.

Bacteria* *(of the harmful variety)* can cause cholera, salmonellosis, shigellosis, and typhoid fever. Bacterial agents can be transmitted in water or food, and in some cases by flies which have been in contact with sewage. Septic tank effluent can contain any of the above pathogens. That's why it's so important for the tank and drainfield to function properly.

Protozoa can cause giardia, amoebiasis, and balantidiasis (causing diarrhea, dysentery, and other intestinal problems). Protozoa agents can be transmitted in water or food, contaminated raw vegetables, and by flies.

Tapeworms, roundworms, and flatworms can cause a variety of problems, usually resulting from direct contact with contaminated sewage or water. They can inhabit the intestinal tract and even migrate throughout the body.

*Bacteria are not well understood in common parlance. Most people associate them with infections, rotting fruit, or smelly socks. Actually, only certain bacteria are harmful, and only if they get into the wrong places. Millions and millions of bacteria reside on or in our bodies, and many of them, such as those that help break down the food in our intestines, are essential for life.

BACTERIA

Bacteria are captured and retained primarily in the biomat (slime layer on trench walls and bottom). When bacteria leave the biomat, they are adsorbed onto (adhere to) soil particles and generally die off over time due to lack of nutrients and proper "host" conditions. Despite high bacteria counts in the biomat, reductions on the order of 1000 times were observed at a distance of only 1 foot into the surrounding soil. At a distance of 2 feet, the counts were in the range for fully treated wastewater. (Hansel & Machmier, 1980.)

If we assume a properly functioning septic system in suitable soil, it's probable that most pathogens will be eliminated in the first few feet of soil. The recommended separations between drainfields and wells (100 feet), drainage ditches (50 feet), cut banks (four times the height of the cut bank, or 50 feet), etc. are obviously well within any safety margin.

In *Septic-Tank Systems: A Consultant's Toolkit,* John H. Timothy Winneberger discusses early engineering studies that showed that community septic tanks removed bacteria from wastewater flows by about 40 to 50%. He goes on to point out that the bacteria going in and the bacteria leaving the tank are probably not the same bacteria, and that our concern should be with specific pathogenic bacteria, not "changes in populations of harmless bacteria."

> In general, pathogenic microbes are rather host-specific and they fare badly outside the host. It is not conceivable that most pathogenic organisms discharged from human hosts would find safety within a septic tank. The environment of a septic tank teems with great varieties of organisms, varying from minuscule viruses to insects and even to higher forms of life at times. Those adapted inhabitants of the septic tank would not likely be without predators to the incoming pathogens.

E. COLI

E. coli (coliform) is one of several bacteria that are normal to the human gut and pass through the intestinal tract with feces. A high coliform count in, say, a creek or pond is often used as evidence of pollution by feces. Further analysis would be required to determine whether the bacteria is actually from humans, and not from wild or domestic animals. The presence of human fecal coliform would certainly indicate a potential human health hazard. (As pointed out above, most pathogens are host-specific, and thus unlikely to infect a different species.)

Scientists have recently identified a rare but dangerous type of the *E. coli* bacterium. Most *E. coli* are harmless inhabitants of the intestinal tract and actually reduce the chance of pathogenic (harmful) bacteria—which enter the body through food and water—from colonizing in the intestines and possibly causing illness.

Because of the knowledge that *E. coli* is a normal intestinal inhabitant, studies of potentially harmful strains were delayed, but in 1982 a strain identified as *Escherichia coli* 0157:H7 was recognized as a variant that produces toxins in the human gut that are capable of deadly damage. Since then, most infections have come from eating undercooked ground beef, but *E. coli* 0157:H7 bacteria can contaminate any food. Undercooked hamburger and roast beef, raw milk, improperly pressed cider, contaminated water, and vegetables grown in cow manure have caused outbreaks of this illness in the United States.

MORE ON VIRUSES

Viruses are electrically charged particles and are removed from the effluent primarily by adsorption onto the surfaces of soil particles. Effectiveness of removal depends upon surface area of the soil particles and effluent flow velocity. When the soil is not saturated, viruses are very effectively removed in the soil. When the soil *is* saturated, removal is not as effective, since viruses will flow with the effluent between soil particles.

Winneberger points out that septic tanks are effective at dealing not only with bacteria, but also with viruses:

> Inasmuch as viruses are charged particles and respond to flocculents,* it seems likely that viruses would become attached to septic tank solids and share their fate in sedimentation. A seeding study was undertaken with a known quantity of polio viruses which was injected into a septic tank. Samples collected from the septic tank outlet and from observation wells in ground waters below disposal trenches contained no viruses.

SOME DEFINITIONS OF SOIL PROCESSES

absorption—The process by which one substance is physically taken into another substance

adsorption—The adhesion of molecules to the surface of solid bodies or liquids with which they are in contact

infiltration—The movement of water into the soil, through a given soil surface

percolation—The movement of water through the soil under the influence of gravity

SOIL AS PURIFIER

As we mentioned before, a healthy aerated soil around the drainfield provides a friendly environment for a rich community of naturally occurring microbes. How does the soil community treat the effluent to remove dangerous pathogens (disease-causing organisms)?

- *Starvation:* It out-competes pathogens for food.
- *Poisoning:* Soil bacteria, fungi, and other microbes produce antibiotics that poison pathogens. (The antibiotic penicillin is derived from a soil mold, penicillium.)
- *Adsorption:* Soil particles have small electrical charges that attract and capture (adsorb) viruses.
- *Filtration:* Larger bacteria can become trapped in spaces between soil particles and filtered out of the effluent.
- *Predation:* The bacteria and many viruses are preyed upon by protozoa. Worms and insects also feed on dissolved nutrients and organic matter (including pathogens).
- *Hostile environment:* The soil environment is so different from that inside the human body that some pathogens simply die from the radically different temperature, acidity, moisture, etc.

SECONDARY OR TERTIARY TREATMENT?

In most wastewater texts, the tank is considered to provide *primary* treatment, the drainfield *secondary* treatment. Engineer Randy May believes that, in actuality, the biomat (*see p. 18*) provides *secondary* treatment and the soil regime *tertiary* treatment.

For technical information on soil types, subsoil aeration, soil drainage, and appropriate treatment processes, see pp. 158–161 in the Appendix.

*A particle of aqueous vapor held in suspension

SOIL LIFE

Under the Microscope

To help you better understand the role of microorganisms in effluent treatment, here are some definitions:

Sun microbe

Microbes: single-cell microorganisms of all kinds

Pathogen: a microorganism capable of producing a disease

Parasite: an organism that lives within or upon another organism, known as the host, without contributing to the host

Bacterium: the smallest of the microbes (except for viruses—see below), about 1/10 the size of a normal cell. *(See footnote, p. 31.)*

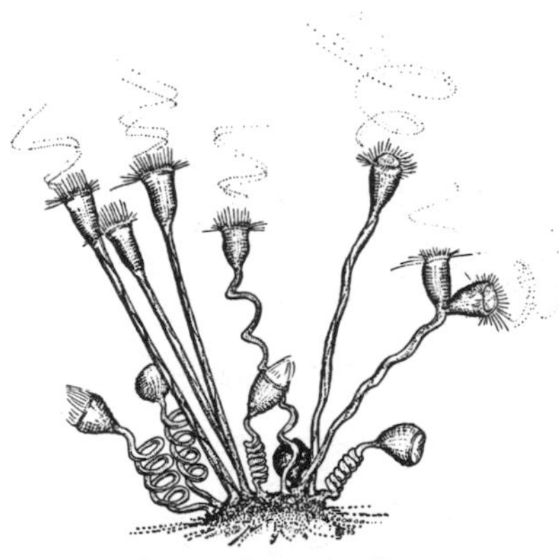

Lolong of Vorticella
A microbe that purifies drinking water

Virus: technically not a microbe, as it is not living in the ordinary sense. A virus is much smaller than a bacterium. "A minute biochemical complex of genetic material, which can take over part of a cell's machinery and use it to produce more viruses."

–*Microbe Power,* Brian J. Ford, Stein & Day: First Scarborough Books Edition, New York, 1978.

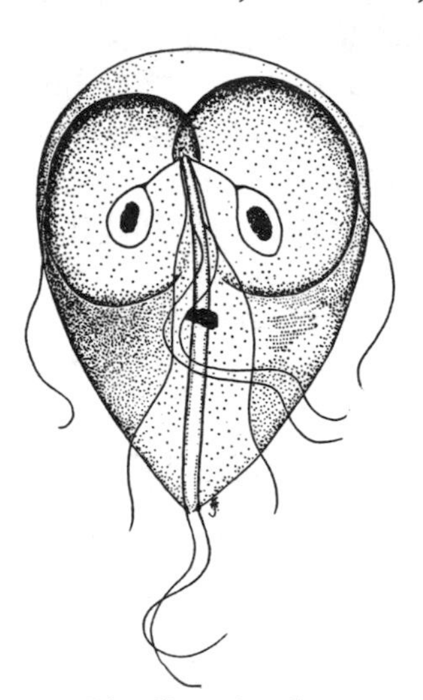

Giardia microbe

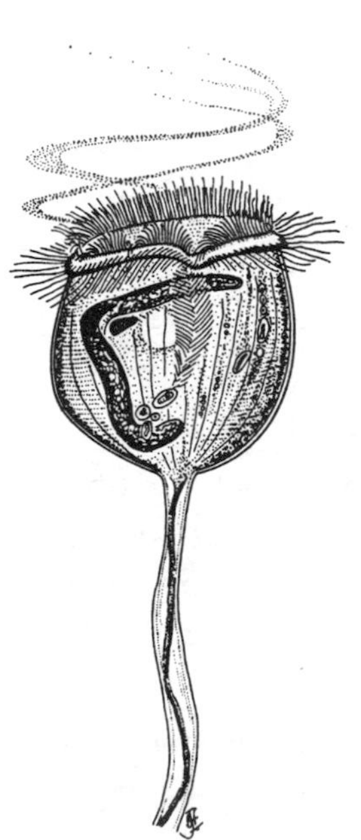

Single Vorticella microbe,
which consumes bacteria

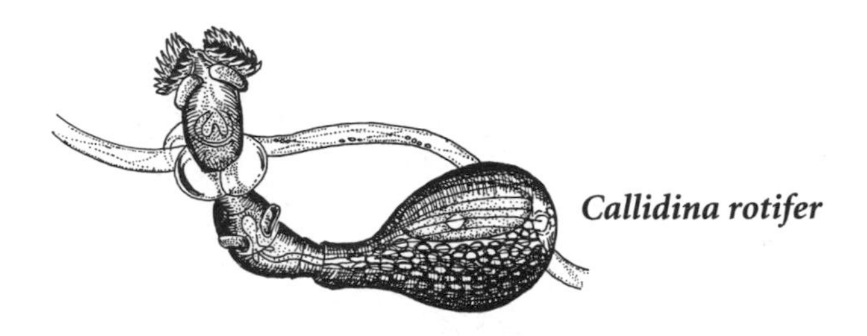

Callidina rotifer

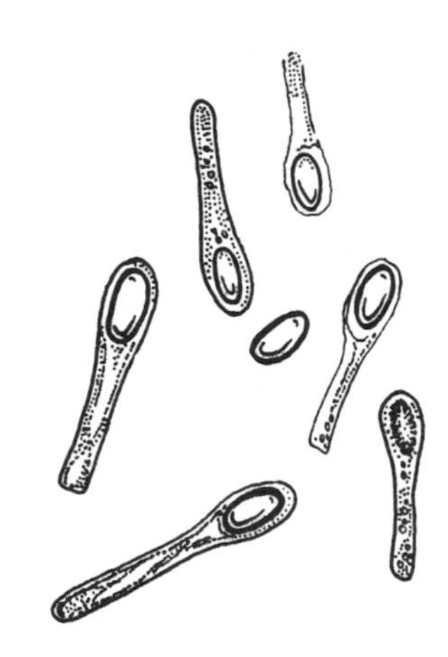

Clostridium bacteria

Rotifers: the simplest and smallest of the macro-invertebrates. They live and swim in water. They stimulate decomposition, increase oxygenation, and recycle mineral nutrients.

Protozoa: microbes, sometimes called "simple animals." Larger than bacteria, they feed on bacteria.

Diffluga amoeba

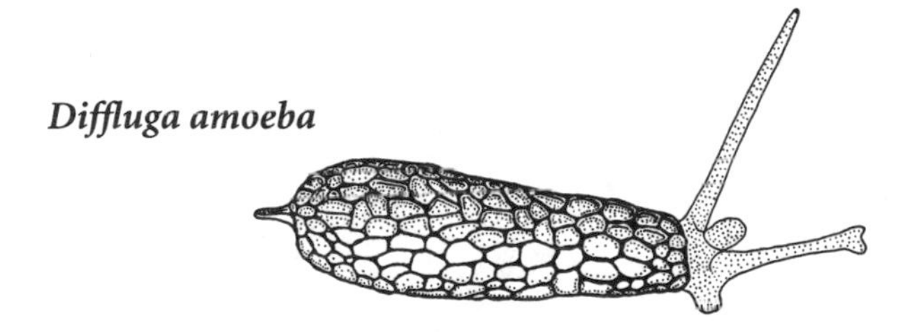

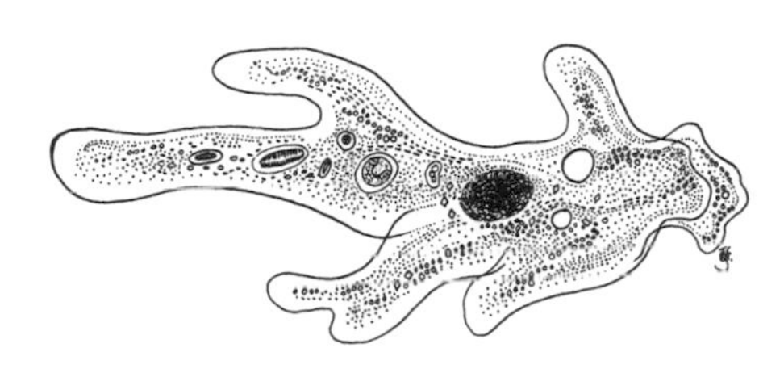

Amoeba proteus

Amoeba: one of many protozoa

Paramecium: one of many protozoa

Algae: a plant-like microbe (carries on photosynthesis). Can be in large chains (like seaweed).

Fungi: plant-like microbes that lack chlorophyll, e.g., mold, rust, mildew, mushrooms, etc.

Note: Most of the above organisms that do survive in the (anaerobic) septic tank do not survive in the (aerobic) soil.

Illustrations by Brian J. Ford from *Microbe Power,* Stein and Day: First Scarborough Books Edition, New York, 1978

"A TEASPOON OF LIVING EARTH CONTAINS FIVE MILLION BACTERIA, TWENTY MILLION FUNGI, ONE MILLION PROTOZOA, AND TWO HUNDRED THOUSAND ALGAE. NO LIVING HUMAN CAN PREDICT WHAT VITAL MIRACLES ARE LOCKED IN THIS LOT OF LIFE, THIS STUPENDOUS RESERVOIR OF GENETIC MATERIALS THAT HAS EVOLVED CONTINUOUSLY SINCE THE DAWN OF LIFE ON EARTH...."

—*CLEAN WATER,* LEONARD STEVENS.
E.P. DUTTON & CO., NY, 1974

4
Down the Drain

CHAPTER FOUR

Down the Drain

No man, woman launderer, or laundresse shall dare to throw out the water or suds of fowle cloathes in open streets, nor within lesse than a quarter of one mile, dare to do the necessities of nature, since by these immodesties, the whole fort may bee choked and poisoned. . . .

–from the first sanitation law in Virginia

DAILY HOUSEHOLD MAINTENANCE

Everyone knows that to keep a household functioning, floors have to be swept, the garbage taken out, and clothes washed. We all know that our cars have to be lubricated and the oil changed periodically. Yet very few people know anything about routine care and maintenance of their septic systems. Many of us grew up in cities, where everything went down the drain, never to be seen or heard from again.

Most septic system problems are due to lack of reasonable care and maintenance. Septic systems do require a few simple steps in daily household management, as well as a timely schedule of inspections and periodic pumping.

In this chapter we'll discuss what you can do on a daily basis to promote a healthy septic system. The previous chapters provided enough of the basics so you'll understand the following recommendations.

WHAT GOES DOWN THE DRAIN

There are three simple, basic things that you can do on a day-to-day basis to help your septic system.

1. Minimize the liquid load.
2. Minimize the solids load.
3. Be careful about what goes down the drain.

MINIMIZE THE LIQUID LOAD

Generally speaking, the less water that goes down the drain, the better your tank and drainfield will work. The more time the anaerobic organisms in the tank and the aerobic organisms in the soil have to digest the wastes, the better job they will do.

THE BIG THREE

The *toilet, washing machine,* and the *bath/shower* are the three biggest sources of household wastewater flow and thus the most important ones to consider in reducing the amount of wastewater the septic system must process. Less wastewater means more retention time in the septic tank and more settling of solids. A reduced flow of cleaner effluent will prolong the life of the drainfield.

Toilet

Old-style toilets use 5 to 7 gallons per flush. Newer ones use 3½ gallons (a 42% savings), and the most recent ones (ultra-low flush) use a scant 1½ gallons (a 75% savings). The ultra-low flush toilets have improved considerably in the past few years. Whereas models in the '70s and early '80s often clogged when handling solids, required double flushing, or were left with "skid marks" in the bowl, the more recent models are better. Most plumbing supply stores now carry these. *(See p. 174 on the Toto.)*

Washing Machine

Connecting the laundry outflow basin to a graywater system is one of the simplest water-saving strategies and can reduce wastewater going into the tank considerably. *(See Chapter 6, "Graywater Systems," p. 64.)*

Bath Water

Re-plumbing the bath/shower to a graywater system can keep 30 to 60 gallons out of the septic system daily. *(See p. 67.)* Installing a low-flow shower head can save 20 to 30 gallons a day.

AVOID SHOCK LOADS

Spread water usage out to maintain adequate retention time and avoid overwhelming the system. Don't do four loads of wash in a row—it's a tremendous volume of water for the system all at once. Don't have three people taking consecutive showers and someone else doing the dishes all at the same time.

OTHER COMMON SENSE WAYS OF SAVING WATER

- Washing dishes: use a small Rubbermaid-type dishpan. Fill with water and wash all dishes in it. Rinse carefully. *(See dishwashing techniques, below.)*
- Take short showers instead of baths and turn shower off when lathering.
- Fix leaky taps. One drip per second from a leaky bathroom tap is 200 gallons per month, 2400 gallons per year.
- Fix a leaky toilet. The culprit here is usually a float valve working improperly.
- Don't let water run while brushing teeth, washing face, cleaning vegetables, rinsing dishes, etc.

THE ART OF WASHING DISHES

This is a dishwashing system worked out over many years. The aim has been to:

- reduce water usage
- reduce fuel used to heat water
- minimize organic matter (oils, rinsed-off food particles, etc.) going down the drain
- save time

To achieve these goals, this operation is performed by hand. No dishwasher (other than the human variety) and very importantly, no garbage disposal. *(See p. 42 on why garbage disposals are tough on septic systems.)*

Sink

The standard American sink, which sits on top of the counter, connected with a chrome lip, is a poor design. Far better is a one-piece sink with drainboards on both sides that drain via gravity into the sink, with no crack or lip in between. Such sinks are available in Europe and some are now available in the U. S.

Hot Water

The kitchen shown here has an insulated 5-gallon electric hot water heater directly under the sink. Propane and natural gas are, in general, better fuels than electricity for heating water (since the energy is generated directly in the home instead of at a distant electric plant), but a small electric heater like this uses surprisingly little electricity when insulated and is quick and convenient to install. Another approach is to use an on-demand heater (only goes on when hot water tap is on).

Is five gallons enough hot water? This system is used by a family of four, with all meals cooked from scratch. It does require hot water consciousness—you can't let hot water continually run when rinsing, for example—but there is generally enough hot water to do the job. When there are guests and a lot of dishes, the heater recovers fast enough to provide additional hot water after, say, 10 minutes.

WATER CONSERVATION

Here's how the system in this kitchen saves water:

- The close proximity of the hot water heater means you have hot water after running less than one cup of water (less wasted water down the drain and less water into the septic system).
- Careful scraping of dishes means much less (or no) use of water to rinse before washing.
- Use of the small rubber dishpan saves water.
- Careful rinsing saves water.

Tools of the Trade

(See p. 42 for illustration.)

- *Scraper:* a very important component of the system; a large one with soft rubber is best. With one of these you can scrape just about all food residue off plates, pots, and pans.
- *Compost bucket:* This one is an excellent design. It has a foot-operated pop-up top and holds a large amount of food scraps. You can dump all liquids in this also (sour milk,vegetable water, fats, etc.).
- *Dishpan:* This is a 12- by 14-inch Rubbermaid container. The sink is not big enough to hold two. (Some people use two dish pans, the second for rinsing, but if soap builds up in the rinse pan, dishes may not get cleansed of soap.)
- *Dish drainers:* two Rubbermaid drainers on the right drainboard with utensil holders and a built-into-the-wall wood dish rack
- *Newspaper:* for wiping grease off frying pans, broiling pans, etc.

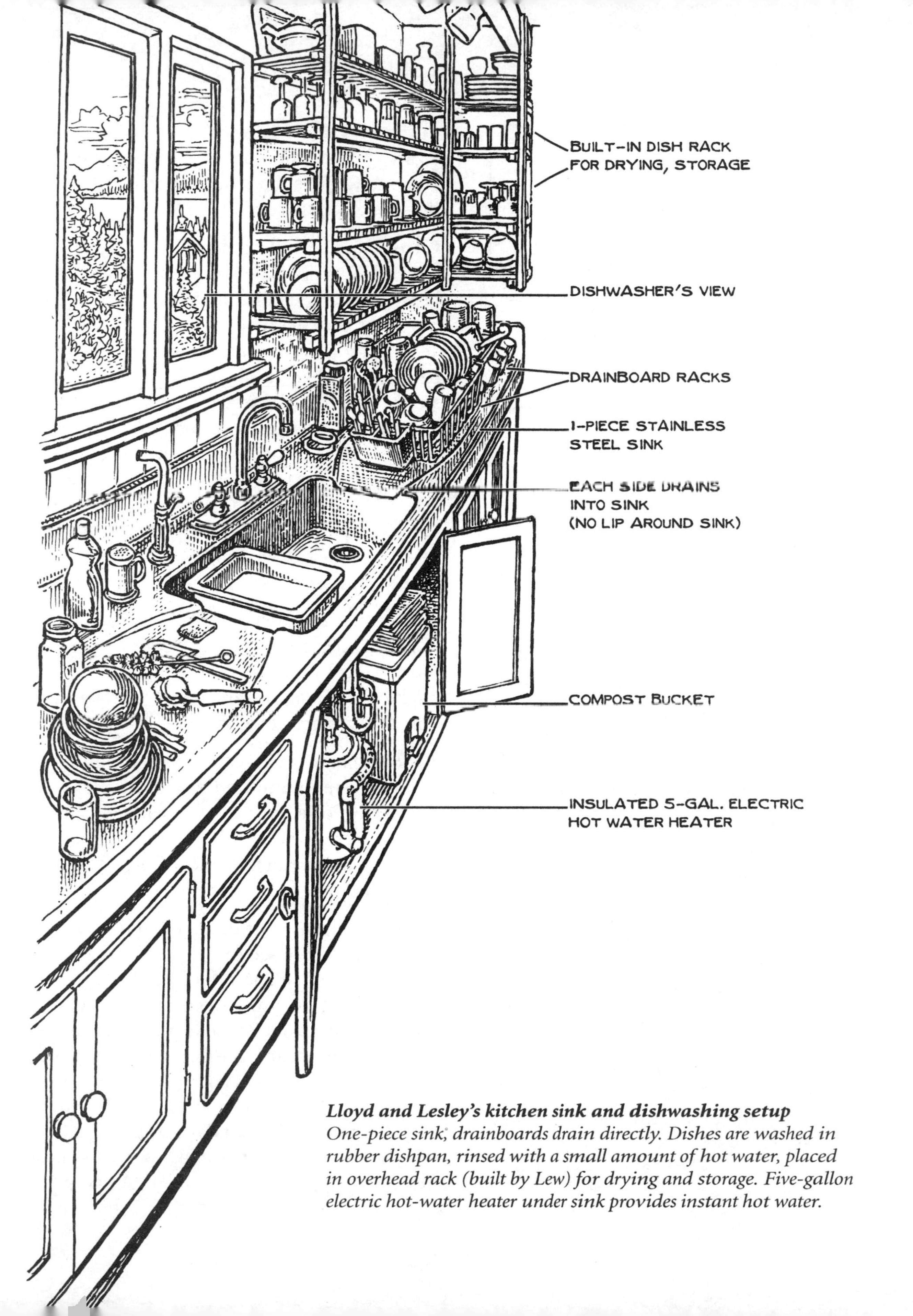

Lloyd and Lesley's kitchen sink and dishwashing setup
One-piece sink, drainboards drain directly. Dishes are washed in rubber dishpan, rinsed with a small amount of hot water, placed in overhead rack (built by Lew) for drying and storage. Five-gallon electric hot-water heater under sink provides instant hot water.

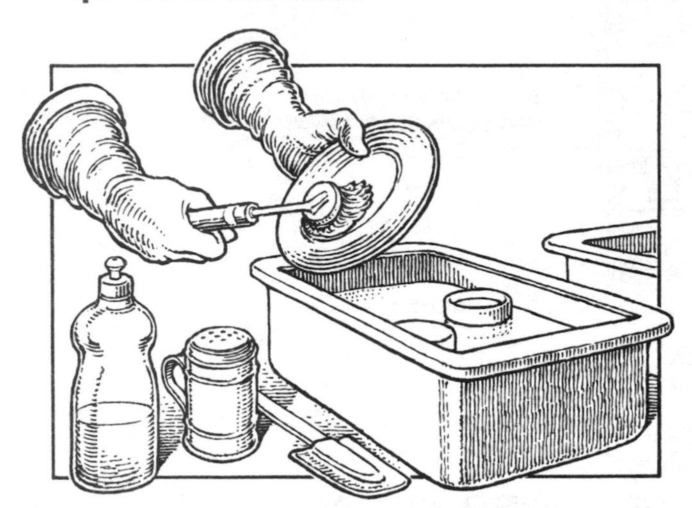

Dishwashing implements
From bottom left, dishwashing soap, baking soda, scraper (essential!), rubber dishpan

Procedure:

- All dishes are scraped (thoroughly) into the compost bucket.
- With greasy pans, newspapers are used to remove most of the grease and put it into the compost.
- Once washed and rinsed, dishes are placed either in the drainboard rack or wall rack.

MINIMIZE THE SOLIDS LOAD

Rule of thumb: Don't use your septic system for anything that can be disposed of in any other way. The less material that goes into the tank, the less frequently it will need pumping.

Garbage disposal: Don't use one if you can help it. Septic system microorganisms work beautifully on solids that have been pre-digested (e.g., feces). Undigested food scraps take much longer to break down and slow down this action considerably. A garbage disposal will increase the amount of solids in the tank by as much 30 to 50%, and, therefore, the tank must be pumped more frequently.

In Chesterfield County, Virginia, any home with a garbage disposal must have it hooked up to a separate 1250-gallon septic tank (in addition to the home's regular tank).

Grease and oil: Collect grease and oil in a container. Do not put them down the drain. Why? First, grease tends to clog drain pipes. (Even if hot water liquefies it, it will cool off and solidify at some point in its journey to the septic tank.) Secondly, the beneficial organisms in your septic system have a hard time with grease and oil; they are not their favorite food. Too much grease can seriously disrupt beneficial septic digestion by microorganisms (as many a rural restaurant has learned to its dismay). First, get all the grease off pans and pots with a spatula, then wipe with newspaper before washing, and put newspaper in the compost bucket.

Paper products: Don't put paper towels or facial tissue (which do not break down as easily as toilet paper) down the toilet. Put no sanitary napkins, tampons, or any paper product other than toilet paper down the toilet.

Synthetic fibers: If your laundry drains into your septic system, minimize the synthetic fibers in your wash. Polyesters and other synthetics do not break down as efficiently as cotton, linen, or wool, and if they exit the tank, will eventually clog leach lines. *(See p. 172 on Septic Protector Filter.)*

Sand and dirt: Anyone or anything that is really dirty should be hosed outside to keep sand and excess dirt out of the drain. Sand is a special problem. A tank inspector in a beach town has found up to a foot of sand in the bottom of tanks. Pumpers can't get this sand out of their trucks, so they are reluctant to pump it from your tank. Someone can go into the tank and shovel it out but this is a **dangerous** practice! A self-contained breathing apparatus is a must. Best not let it get into the tank in the first place.

A DAILY WATER BUDGET

Water conservation practices ☐ compared with normal water use ☐

ACTIVITY (PER PERSON)	CIRCUMSTANCES	WATER USED	TOTAL
TOILET 4 flushes per day	ultra-low flush toilet	1.6 gallons per flush	6.4 gallons*
	conventional toilet	3.5–7 gallons per flush	14–28 gallons
SHOWER Once a day, 5 minutes	low-flow showerhead	2.5 gallons per minute	12.5 gallons*
	conventional head	3–8 gallons per minute	15–40 gallons
BATH Once a day	tub ¼ to ⅓ full	9–12 gallons	9–12 gallons
	full tub	36 gallons	36 gallons
SHAVING Once a day	one full basin	1 gallon	1 gallon*
	open tap	5–10 gallons	5–10 gallons
BRUSHING TEETH Twice a day	wet brush and rinse	¼ to ½ gallon	less than 1 gallon*
	open tap	2–5 gallons	4–10 gallons
WASHING HANDS Twice a day	one full basin	1 gallon	2 gallons*
	open tap	2 gallons	4 gallons
COOKING** Washing produce	1 full kitchen basin	1–2 gallons	1–2 gallons*
	open tap	5–10 gallons	5–10 gallons
AUTOMATIC DISHWASHER Once a day, full load	short cycle	8–12 gallons	8–12 gallons
	standard cycle	10–15 gallons	10–15 gallons
MANUAL DISHWASHING Once a day	full basins, wash/rinse	5 gallons	5 gallons*
	open tap	30 gallons	30 gallons
LAUNDRY Two full loads per week	portion of full load	10–15 gallons	10–15 gallons*
	full load	35–50 gallons	35–50 gallons
CAR WASHING Twice a month	5 full 2-gallon buckets	20 gallons per month	⅔ gallon/day
	hose, shut-off nozzle	100 gallons per month	3.5 gallons/day
LAWN TREES, SHRUBS, GARDEN	Watering requirements vary with plant species, type of turf, season, region, and soil type.		

*Total with water conservation practices = about 50 gallons per day per person.
**Real cooking figures will be higher to include boiling water, rinsing utensils, etc.

WATCH WHAT GOES DOWN THE DRAIN

We do not recommend putting any of the following down any house drain:

- Drano, Liquid Plumber, or any lye-based chemicals
- Oven-Off or any other strong cleaning agents
- Oil, paint thinner, or photo lab chemicals (although home photo lab rinse waters are considered fairly safe)
- Root Deterrent or any product containing copper sulfate. Commonly sold in hardware stores, these products may kill clogging roots, but by the same token will kill the beneficial living microorganisms in your system.
- *Septic system additives*, especially enzymes. (You don't need to add enzymes; they're naturally present in the sewage.) Beware of telemarketers or ads hawking additives claiming to avoid tank pumping. They actually break down the scum and sludge into small particles, which are then readily flushed out into the drainfield, increasing possibility of premature drainfield failure. The State of Washington has banned septic tank additives. In Tiburon, California, a homeowner recently added enzymes to a septic system that had been working perfectly well. Soon after, sludge moved out into the drainfield and the system failed.

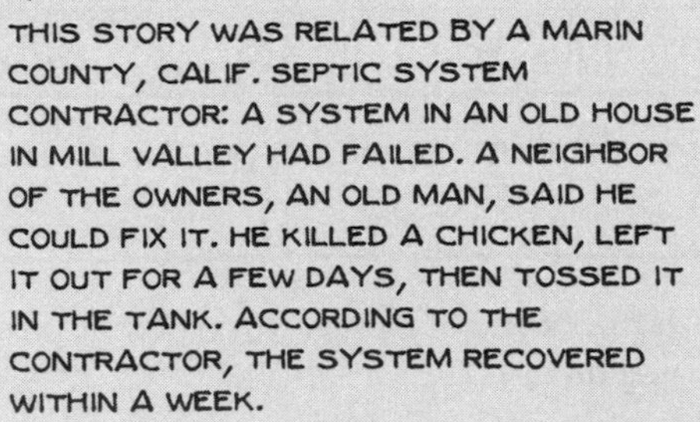

Responding to the belief held by many septic tank owners that adding cakes of baking yeast is beneficial, John H. Timothy Winneberger, Ph.D., a well-known septic systems pioneer, said the only way yeast could benefit a septic tank is if "you eat it first."

- Excessive chlorine bleach and toilet additives. Products such as Thousand Flushes and BluBoy are mainly chlorine, which is noted as being unfriendly to microorganisms in the system.

SEPTIC LIFE IS HARD TO KILL

Recent studies, however, have indicated that household chemicals may not do as much damage as was assumed, and that the bacterial population of tanks can recover quickly. An article in a 1997 issue of *Small Flows* states:

> Research conducted over the past several years has concluded that with normal use, household cleaning products do not adversely affect septic tank operation. Normal use of household cleaning products is considered to be the amount recommended by the manufacturer. . . .
>
> A study conducted at the University of Arkansas at Little Rock determined the amount of household chemicals required to destroy all bacteria in an individual domestic septic tank. Domestic cleansers, disinfectants, and drain openers were used in the test.
>
> Results showed that an excessive amount of any one of the cleansers or disinfectants applied in a slug loading (all at once) was required to destroy the bacteria in the septic tank. However, after normal septic system usage, the bacterial population recovered to its original concentration within hours. In other words, although under extreme stress and shock loading conditions, the bacteria can be destroyed, rejuvenation does occur within hours.

What this study did not address was daily use of strong detergents and/or other chemical products, which could hinder healthy bacterial life in the tank. It's also a fact that many people consciously avoid buying toxic products due to by-products created in their manufacture.

DRAIN CLEANERS ARE THE WORST

The above study also indicated that even small amounts of drain openers can kill the entire bacterial population in a tank. A more sensible (and elegant) solution to unclogging drains is the Drain King, which uses only water and pressure. *(See p. 174.)*

WATER SOFTENERS BAD FOR TANKS

Water softener backwash contains high levels of chlorides, which not only can kill microorganisms in the tank, but also will interfere with sedimentation of solids. It should not be routed into the tank.

NOTE TO CITY FOLKS

ALTHOUGH THE FOCUS OF THIS BOOK IS ON WASTEWATER DISPOSAL IN RURAL AREAS AND THE SUBURBS, YOU CAN APPLY THE SAME PRINCIPLES LISTED IN THIS CHAPTER IN CITIES AND YOU'LL NOT ONLY SAVE WATER, BUT CONTRIBUTE TO MORE EFFICIENT WASTEWATER DISPOSAL AND PROTECTION OF GROUNDWATER, LOCAL STREAMS, LAKES, OR THE OCEAN.

SUMMARY

To sum it up, there are many steps you can take on a day-to-day basis to promote a healthy septic system and to prevent system failure:

- Install a low-flush toilet and/or a low-flow shower head.
- Use a graywater system for laundry and/or bath wastewater.
- Don't allow taps to run. Fix leaky taps.
- Wash dishes economically.
- Don't use a garbage disposal. Compost your kitchen waste if possible.
- Don't put anything other than toilet paper down the toilet drain.
- Don't put caustic products, grease, oils, or septic tank additives down the drain.
- Minimize bleach and other chlorine products.
- Don't put unnecessary dirt or sand down the drain.
- Use biodegradable soaps.

FEAR NOT!
Call
84

5 Septic System Maintenance

CHAPTER FIVE

Septic System Maintenance

In the last chapter we talked about what goes down the drain. Here we're going to cover long-term periodic maintenance, which consists mainly of septic tank inspection and pumping when necessary. We'll also discuss drainfield inspection.

People often say, "Oh, I've never had to pump my tank," as if that were proof that their septic system works fine. But be aware, failure to pump tanks is (next to improper siting and design) perhaps the greatest single cause of septic system failure. Here's what can happen:

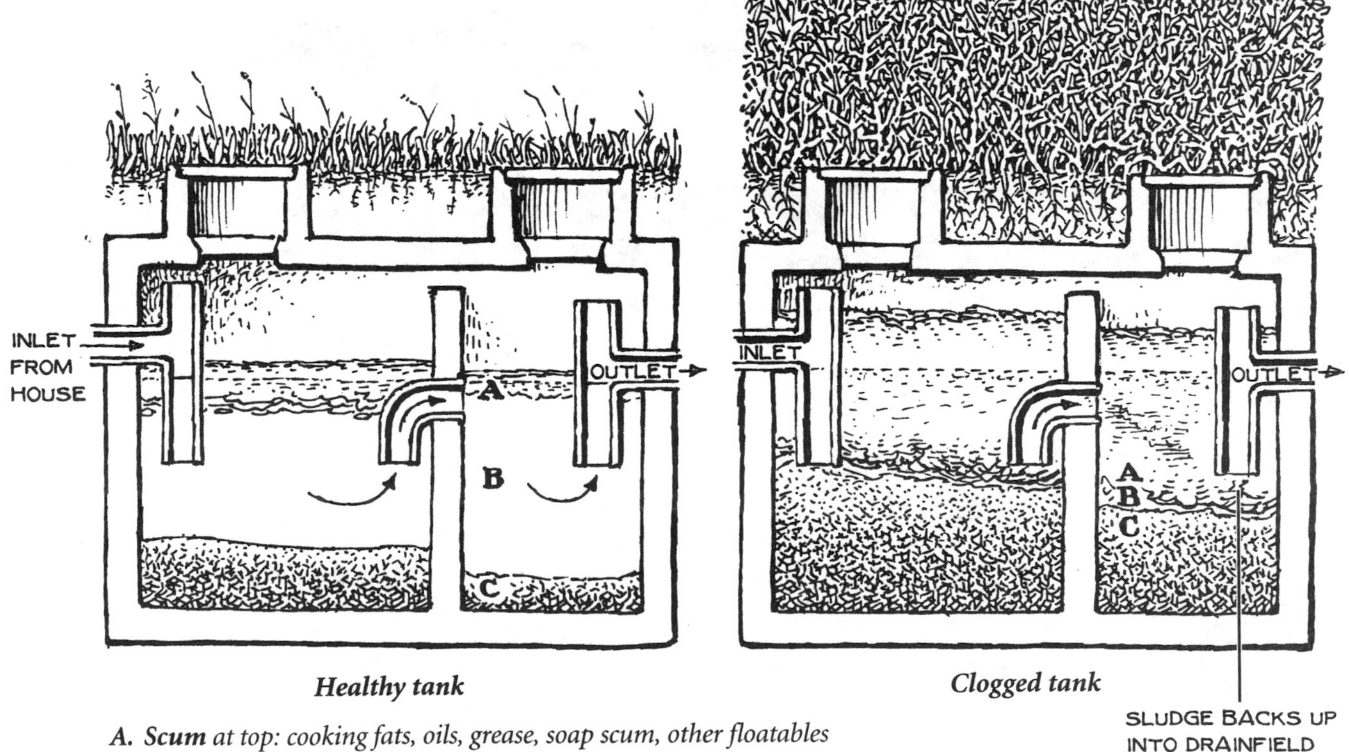

Healthy tank ***Clogged tank***

A. Scum *at top: cooking fats, oils, grease, soap scum, other floatables*

B. Liquids *in middle*

C. Sludge *at bottom: solids heavier than water and what is left over after solids have been partially eaten by bacteria. Once sludge gets up to outlet pipe, it enters and clogs drainfield.*

DRAINFIELD FAILURE

After several years of use, a build-up of bottom sludge and floating scum will reduce the effective capacity of the tank, as shown in the "clogged tank" illustration on page 48. This means waste passes through the tank too fast, and solids eventually plug the pipes in the drainfield. The microorganisms in the drainfield no longer have an aerobic (with air) environment in which to perform their cleansing action; they are now struggling to survive in an anaerobic (without air) environment. Either untreated effluent begins surfacing on the ground or sewage backs up into house drains. At this point, the system has *failed*, and a new drainfield is required—expensive!

THE MOST IMPORTANT SINGLE THING A HOMEOWNER CAN DO IS TO AVOID PLUGGING OF THE DRAINFIELD. AN EXCELLENT AID IN DOING THIS IS AN EFFLUENT FILTER *(SEE P. 9 AND P. 174)*.

INSPECTION AND PUMPING

Inspect the Tank

How can you avoid drainfield failure? Inspect the tank at regular intervals and pump when necessary. In many parts of the country, it is recommended that tanks be pumped every three to five years, but recent studies indicate that a functioning tank, without abuse, may only need pumping every 10 to 12 years. Since there are many variables, we recommend an inspection every three to five years and *basing pump-outs on inspections.* As the years pass, you should be able to see the pattern of sludge and scum accumulation.

Keep a Record

Use a file folder (or get your wastewater district to get the *Homeowner's Septic System Guide* shown on page 180) to keep a record of inspections and dates when the tank has been pumped.)

What Is Pumping?

Septic tanks are pumped by a licensed pumper with a vacuum tank truck. The pumper will use a 4-to-6-inch-diameter hose and vacuum everything out of the tank (both solids and liquids).

Waste pumped from a septic tank is called septage. It is approximately 5% solids and 95% water. (Raw sewage is 1% solids and 99% water.) The septage waste must be taken to a licensed disposal site because of the potential health problems with contamination. In many rural areas, private companies have developed septage disposal sites—generally evaporation ponds. In other communities, there may be a centrally located sewage plant that can handle the septage waste.

WHERE IS IT?

Locating the Tank

You can save some money by locating the tank yourself and digging up the manhole covers. If the tank has no risers over inspection holes, and no diagram is available showing the location, you will have to probe for the tank, as follows: Use a long metal rod (½-inch rebar, bent over 90° to make a handle at the top) and begin probing where the main drain pipe leaves the house. Push the rod firmly down into the soil until you "feel" the drain pipe. Use a firm and steady push. Don't punch or pound the rod as you can damage the pipe, particularly the pipe/septic tank connection. If the soil is too hard and dry for probing, try soaking the area with a garden hose.

Another method: There may be lush growth over the drainfield. Then the tank will be in an obvious place between the house drain and the drainfield. Or, you can run a snake down the clean-out to the tank and locate it with a metal detector.

When you find the drain pipe at one spot, move a little further from the house and probe again. Continue along the path of the drain pipe until you locate the tank. The tank will probably be 1 to 3 feet underground and at least 5 feet from the building. Once you locate it, dig up both manhole covers. Or, if you're lucky, the tank will have risers with sealed caps instead of the very heavy manhole covers of earlier models. If you plan to inspect your own system and don't have these risers *(see p. 6)*, we recommend that you have them installed. In addition to providing easy access for inspection, they keep out dirt and rainwater. In the meantime, use a rope through the metal handles on the concrete manhole covers to swing them up and off the tank. The tank is now ready for inspection and/or pumping.

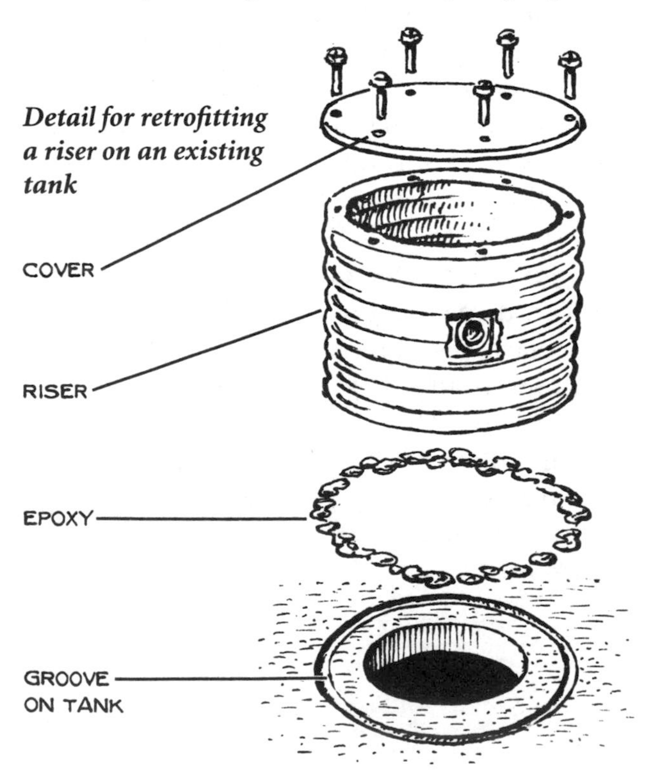

Detail for retrofitting a riser on an existing tank

Note: Once you locate your tank, make a diagram of where it is for future reference, indicating number of feet from a particular point of the house.

Know Where Thy Tank Lieth!

B & M Contractors, of Bolinas, Calif., tells the story of some people who added a kids' bedroom to their house without checking the location of the septic tank. All went well until one day, the system failed and it was then discovered that the room had been built over the the tank. To get to it, the pumpers had to pull back the rug, cut a hole in the floor, run their suction hose into the room through a window, and pump out the tank. Yuck!

TANK INSPECTION

Checking It Out

You can save money by doing your own inspections. This way you will only call the pumper when needed. Inspection is done from above, by looking in through the manholes. Look around inside with a flashlight and perhaps even a hand mirror attached to a long pole. When checking tanks be sure to wear gloves and to wash your hands thoroughly with an antibacterial soap afterwards.

However, if you've had no experience, it's hard to know what to look for. If you intend to make your own inspections, we suggest you have the pumper come out the first time and that you watch how s/he performs the inspection. Ask questions. Then, the next time you should be able to do it yourself.

> WARNING: BE VERY CAREFUL AROUND SEPTIC TANKS. THE FUMES CAN KNOCK YOU OUT. FALLING INTO A TANK CAN BE LETHAL. NEVER ENTER A TANK. FOR A SUMMARY OF SEPTIC SAFETY, SEE WWW.INSPECT-NY.COM/SEPTIC/SEPTICSAFETY.HTM .

What to Look For

Once the tank is open, here's what to look for (assuming the tank has two compartments):

Inlet Chamber

1. *Odor:* Odors should not be *too* obnoxious when you open the inlet side. (Odors will be a lot stronger when you stir the contents.)
2. *Insects:* There should not be too many flies or flying insects present.
3. *Scum:* Should be firm, with a crust, but not solid. It should be like pudding, a medium brown color, and 3 to 4 inches deep. By poking a stick through the scum, you can estimate the average thickness. Or, you can fashion an "L-rod," as shown at the top right. You can figure on there being equal amounts of scum above and below the water line.)

   Tip: Sometimes you can use a hose with high pressure to squirt a hole in the scum big enough to estimate its thickness.
4. *Sludge:* You can use a long stick, but best is a concrete hoe (the type with two holes is best) and an extension handle wired or taped on. As you lower the hoe, it's a little tricky is to tell when you first hit the sludge. Thus, proceed slowly. If you feel resistance halfway to the bottom, it needs pumping.
5. *Inlet tee:* Concrete tees deteriorate. Be sure to check this.

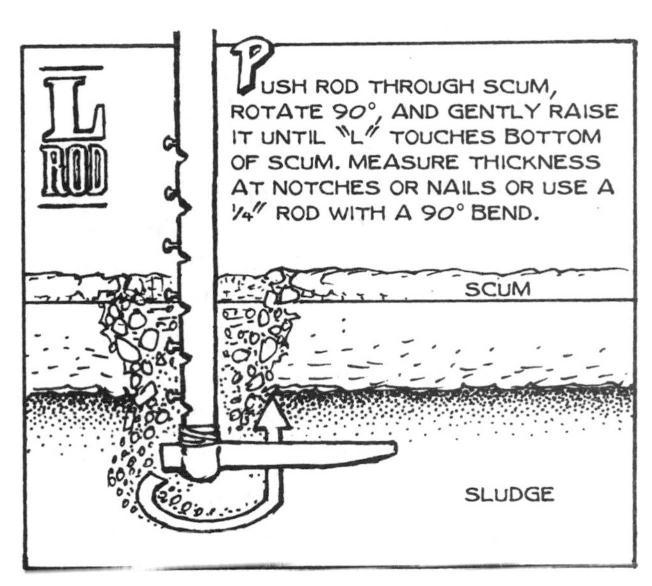

Homemade scum-measuring device

A COOL SLUDGE/SCUM TOOL

GET A 1-INCH CLEAR PLASTIC TUBE, 5 TO 6 FEET LONG. SLOWLY PUSH THE TUBE TO THE BOTTOM OF THE TANK, THEN COVER THE TOP WITH YOUR THUMB AND REMOVE CAREFULLY. WIPE THE TUBE OFF, AND YOU SHOULD BE ABLE TO SEE A PROFILE OF YOUR TANK, INCLUDING SLUDGE, CLEAR EFFLUENT, AND SCUM.

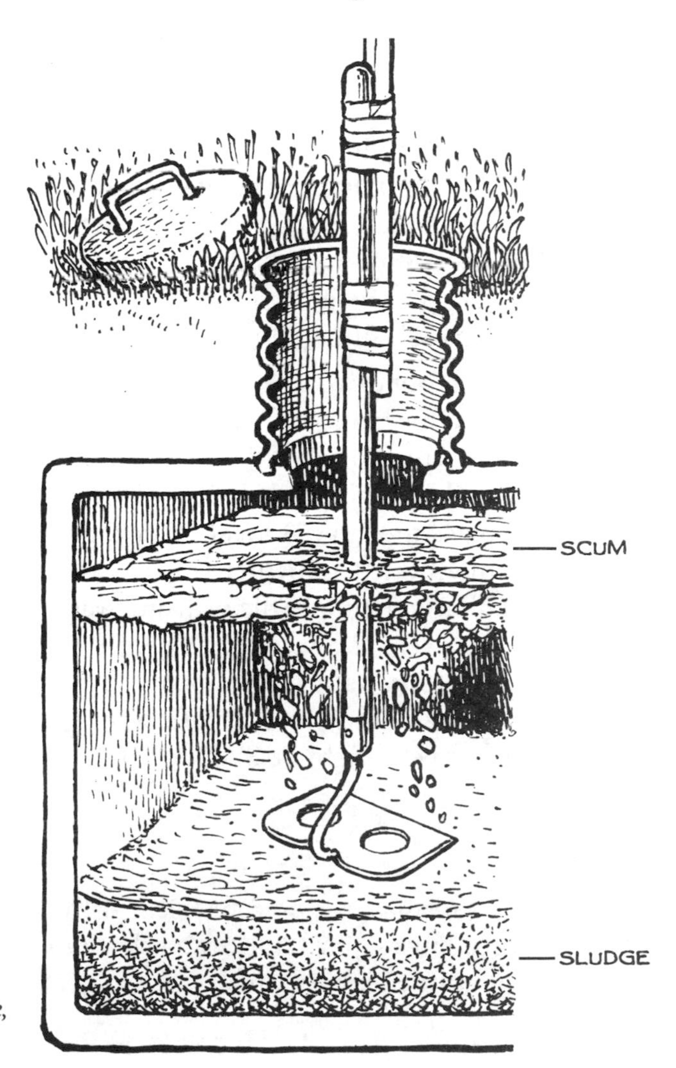

Homemade sludge-measuring device
You can use hoe to measure depth to sludge, then a rod to measure depth of sludge.

Outlet Chamber

- *Scum:* In a *two-compartment tank*, there should be little, if any, scum on the effluent side—the effluent should appear relatively clear. If there is much scum here (more than 2 inches), the tank needs pumping. If either scum or sludge is floating out the outlet, the tank needs pumping. In a *one-compartment tank*, a rule of thumb is that the tank should be pumped when the sludge is 20 inches and the scum is 10 inches.
- *Outlet tee:* If the inside of this tee is clogged, the tank is flooding, and this could indicate trouble with the drainfield. If the top is dry, it's a good sign, since a wet top would indicate the tank is flooding. If the tank is flooding, there is no air at the top of the tank, and this anaerobic condition can result in tank deterioration.
- *Outlet tee deterioration:* A concrete or ceramic outlet tee in a tank can deteriorate above the water line due to sulfuric acid. This is easy to replace with a plastic tee and should be inspected periodically.
- *Baffle wall deterioration:* The baffle wall between the two chambers can deteriorate as well. Consider putting an effluent screen in place rather than trying to repair the baffle wall, or replace the tank. A local septic tank inspector mentioned an owner who went into his tank to repair a baffle wall and was sick for over a year as a result. In Oregon, for example, most new tanks are now one chamber with an effluent screen. Cost for installing a screen might be $200 to $300. *(See pp. 8–9.)*

Insects

Mosquitoes and flies can be a problem if they enter and breed in a septic tank. Strangely, this is not often mentioned in literature on the subject. Mosquitoes and flies can enter through the plumbing vent of the house, go down through the 4-inch drain pipe and through the inlet tee to the tank. They can then breed in the tank and travel via the same route, reversed, to the outside world. You can cover the top of the vent with a capper of stainless steel screen. Another place for mosquito entry can be tanks with wood or fiberglass risers; here the manhole covers can be sealed with roof patch or a plastic sheet over the lids, then covered with a few shovelfuls of sand.

IF THE TANK NEEDS PUMPING

Try to be there when the pumping is done. Lean over the shoulder of the pumper and make sure the tank is pumped completely. We heard about one company that pumped only the liquids and no solids. As the tank is pumped, it should be cleaned out as thoroughly as possible with a hose. There will be plenty of bacteria left to reactivate the system even when the tank is thoroughly cleaned.

It *is* difficult to suck out the bottom 2 to 3 inches of sludge, particularly if it contains a lot of sand. The pumper should hose down the sludge on the bottom when it is exposed so that it will partially liquefy and can then be sucked out. A high-pressure squirter, not a thumb applied to the hose, should be used.

DRAINFIELD INSPECTION

If the drainfield was properly designed and installed (and the tank functions properly), it should be mostly maintenance-free. However, here are some tips.

Drainfield Test

Lush plant growth over the drainfields (or tank) may be a sign of sewage surfacing. Here's one way to check the drainfield's absorption capacity: run 40 to 80 gallons of water into the

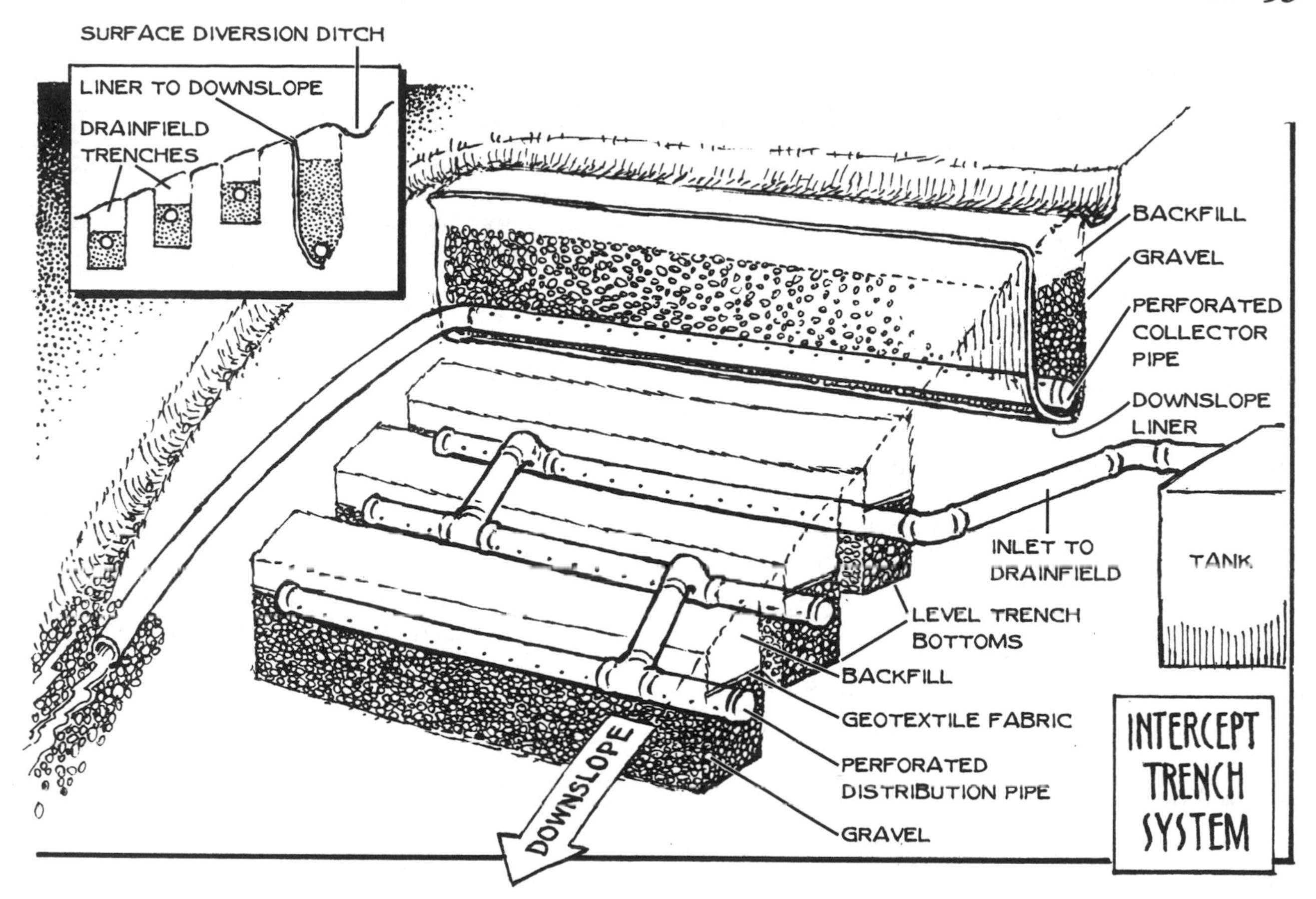

tank and, with the outlet riser open, watch how long it takes to drain into the drainfield. A slight back-up from the normal tank water level (the bottom of outlet pipe) is OK. The water level may rise ½ to 1 inch, but if the field is unclogged (and not flooded), a fully-flooded tank should drop to normal in five to ten minutes.

The "French Drain"

If the original installation didn't allow for good drainage of surface waters (rain or run-off) around the drainfield, drainage ditches (with proper setbacks from the drainfield) may be needed. Also, a high water table in winter and/or dense soil can cause effluent to surface, causing bad odors and a possible health hazard.

Roots

Trees or shrubs with aggressive, water-seeking roots growing near the drainfield can cause real problems in conventional drainfields. The roots will seek out water, and can run inside the drainfield pipes and choke off the flow of effluent. Willow roots are notorious drainfield invaders.

Dual Drainfields

If you have a dual drainfield with a diverter valve, rotate the valve to the alternate field every six months or year. *(See p. 20.)* This allows the trenches to dry out and rejuvenate.

Soil Compaction

Be sure no one parks cars over the drainfield. It will compact the soil and reduce the aerobic capacity of the drainfield. Also, be sure cars don't drive over the inlet and outlet pipes to the septic tank. This can snap the pipes and even cause the tank itself to crack.

Pump Maintenance

For mounds, sand filters, STEP systems, and lift systems for gravity drainage fields:

- Run the pump through its cycle periodically to make sure there are no leaking pipes.
- Pumps should have alarms.
- Pumps should have check valves. *Note:* in areas of severely cold weather, check valves can be detrimental to proper pump operation.

Keep Those Hands Clean!

Just as you can pick up a cold or the flu by getting germs on your hands and then touching your hands to your nose, you can pick up some much nastier organisms if you have manual contact with an open tank or drainfield. If you're going to do your own inspection or repairs, wear gloves and wash your hands scrupulously afterwards.

Never Been Pumped

The fallacy: You often hear a homeowner say, "Oh, I've never had to pump my tank."

The irony: Typically, the homeowner has never had any septic system problems and thinks this means nothing need be done.

The remedy: Just as you need to check the oil level in your car so it doesn't get too low, you need to periodically check the solids level in your tank so it doesn't get too high.

POLITICAL PUMPING

ONE HOMEOWNER IN OLYMPIA, WASHINGTON HAS A SIMPLE REMINDER TO PUMP HIS TANK. HE PUMPS IT OUT EVERY YEAR THAT THERE IS A PRESIDENTIAL ELECTION.

SUMMARY

- Sludge and scum accumulate in every septic tank. How much and how fast depend upon a number of conditions.
- You can save money by locating the tank yourself and inspecting scum and sludge levels to determine when pumping is necessary. (However, there is a learning curve.)
- Inspect your tank every three to five years until you determine the inspection frequency required for the future.
- Scum and sludge must be pumped on a regular basis or system failure can result.
- Drainfield operation can be maximized by checking the tank's outlet tee, by testing the drainfield's absorptive capacity *(see pp. 52–53)*, by ensuring good drainage, and by avoiding soil compaction.

OUT OF SIGHT, BUT IN YOUR MIND!

6
Red Alert!....
System Failure

CHAPTER SIX

Red Alert! . . . System Failure

Your system has failed—that's presumably why you're reading this chapter. Water has backed up into the shower, the toilets won't flush, and/or drains won't drain. This means wastewater has backed up from the tank through the main drain into the house. *It's going the wrong way!* Or—untreated effluent is surfacing on the ground. In this chapter we'll talk about different types of failures, their causes, and what to do when your system fails.

PROBABLE CAUSES OF FAILURE

One or more of the following may have happened:

1. The sewage pipe between the house and the tank is blocked or broken.
2. Either the inlet or outlet tee is blocked or broken.
3. The line between the tank and drainfield is blocked or broken.
4. If the system has a pump, there may have been an electrical or mechanical failure.
5. The tank itself is blocked with solids or has collapsed (an old redwood tank perhaps).
6. The drainfield is flooded due to heavy rains or flooding.
7. The drainfield is (partially or completely) clogged with solids or roots.

LOCATING THE PROBLEM

When the septic system is failing, there is a procedure for locating the cause, called the *discovery process,* in which you search for the problem in the following order:

System Blockage

You start by searching for a blockage somewhere in the system because this is the easiest cause to locate, and the easiest (and cheapest) problem to solve.

1. If only one fixture does not drain, check for blockage between the fixture and the main drain pipe. Use clean-outs for checking.
2. If all fixtures on one branch of the drain pipe do not drain, check for blockage in that branch. Also, check the tank inlet for blockage.
3. Open the tank.

   If it is flooded, the problem may be at the outlet or beyond. Check the outlet for blockage.

   If it is not flooded, you can check the various household fixtures by running a hose down them to see if the water makes it to the tank.
4. If sewage is not arriving at the tank, then check for a pipe line blockage between the house and the tank.

5. If both the outlet and the inlet tees are good, but sewage is still backing up in the tank and house plumbing, the problem may be in the tightline (pipe between tank and drainfield) or the drainfield itself.
6. To check the tightline, you'll have to dig it up where it enters the drainfield. Try a plumber's snake to check for a blockage between the tank and drainfield. Or, the tightline may be broken or sheared off. *(See at right.)*
7. If the tightline is clear and intact, and all of the above steps have not produced the culprit, the drainfield is probably the problem. *(See next page.)*

Clearing Pipe Blockage

If the plumbing suddenly backs up under normal use, especially in dry weather, blockage is the prime suspect. This is generally the easiest problem to correct, particularly if it's between the house and tank.

Most pipe blockages can be located using a plumber's snake. (All tool rental stores have snakes.) Or, an old garden hose may work if there are not too many bends in the pipe. Also, there is a simple (and brilliant!) rubber device called the Drain King, which fits on the end of a garden hose. The hose is then run down the clogged drain pipe, and when the water is turned on, the bulbous rubber section expands, locks in the pipe, and emits strong pulsating bursts of water. These are available for 1- to 10-inch drains. *(See Appendix, p. 174 for more information on the Drain King.)*

Root Blockage

If you find that roots between house and tank are the problem, a Roto-Rooter can clear the line, but the roots will return if the entry points (leaks) are not found and sealed.

Tightline Breakage

A common problem is that the tightline (pipe between tank and drainfield) has broken. This often happens when the tank, which is very heavy when filled, has settled in the ground some time after installation, and the pipe has not flexed. (In some areas, new systems now must include a flexible coupling at the septic tank wall.) As with pipe blockage, snaking the line usually helps you find this problem so the pipe can be either repaired or replaced. Sometimes it is easier and cheaper to replace the tightline, especially if roots are the problem, than to try to clear it.

Power Outage/Flooding

If you have an alternative system that uses electricity to pump the effluent to the drainfield and there is a power outage, the pump chamber should be checked immediately. If it is low, you're OK, but monitor your water usage. If it is nearly full, water usage must be severely curtailed or the result will be effluent from the pump chamber backflowing into your house since there's no electricity to pump the effluent to the drainfield—a distinct disadvantage to high-tech systems! You'll have to wait until the electricity comes back on, or keep a small standby generator on hand for such emergencies.

If you live in an area with high groundwater and heavy rains, your tank and your pump chamber might be filling with rainwater runoff. It is wise to have good risers around the inspection holes of the septic tank since concrete lids generally leak. These risers are sold in many builders' supply outlets. *(See p. 50.)*

TANK FAILURE

Every tank has an estimated life span.

Redwood tanks: They can last from 15 to 45 years. Redwood tanks as well as steel tanks are now illegal in many areas. The top and upper sidewalls of a redwood tank will deteriorate first, causing the top to cave in (this can be dangerous, especially for kids) and the tank will start leaking in the upper 6 inches of liquid level. If a redwood tank fails completely, it's best to replace it with a concrete or fiberglass one. However, if the sidewalls and bottom are good, installing a new top may gain you another 10 years.

Fiberglass tanks: They are light to handle, and quite durable, but they have been known to break due to shifting ground and are hard to repair.

Polyethylene tanks: They sometimes lack structural integrity. Manufacturers claim an estimated life span of 30 years.

Concrete tanks: They are usually precast and are the best and most common type of tanks.

- Concrete tanks can develop leaks if the ground shifts, if the concrete was of poor quality, or (*very* common), when the joint between top and body leaks. Further:
- The tank may be porous and leak from the start.
- The tank may not be properly sealed and small leaks can grow larger. Leaks can be fixed, but the tank must be properly cleaned and prepared, and techniques are difficult and dangerous in an older tank.

All tanks can leak around the inlet and outlet connections. This can usually be repaired by resealing the joints with caulk or mortar. Due to accumulation of gases, concrete outlet connections often disintegrate. Old ones should be replaced with plastic fittings.

Warning: Any underground tank that has had sewage in it is dangerous due to gases, especially if you are working alone. Only a trained professional with a self-contained breathing apparatus should enter a septic tank. Repairs to inlet and outlet tees can usually be completed without entering the tank, and should be done when the tank is pumped.

DRAINFIELD FAILURE

This is the most serious and costly type of failure. Pipe blockages can be removed. Loose connections can be fixed. A faulty tank can usually be repaired. But if the drainfield is clogged, it must be replaced, a disrupting and costly procedure.

Suspicious Drainfield Behavior

You can suspect the problem is the drainfield if:

- there are odors or persistent wet spots over the field
- the plumbing becomes sluggish over a period of time, or when it's used heavily, or during wet months
- problems persist even though the tank has been pumped recently
- the septic tank is flooded

Why Do Drainfields Fail?

- **Clogging with solids:** When a tank has not been pumped periodically, it fills up and eventually the solids migrate out of the tank into the drainfield. Also, old drainfields can be clogged by soil infiltration. Either the perforated distribution pipe or the pores in the soil become clogged.

- **Root blockage:** Root growth near a septic system is a mixed blessing. The good news is that plant growth over a drainfield will absorb much of the discharge; further, in arid areas, evapotranspiration will release the water back into the environment. The down side is that the same root growth that absorbs water can clog disposal lines and trenches and hinder drainfield function (although root blockage is not the same problem in a shallow drainfield).

  Root deterrent products do kill roots, but we don't recommend them, any more than we'd recommend putting Drano down the drain. Most of these products contain copper sulfate, or "bluestone," which can kill off the beneficial organisms in both the tank and the drainfield, and poison things if it gets in the water table. (These products also produce toxins in the manufacturing process.) Far better to snake out the line and then remove the source of the roots.
- **High groundwater**: When a drainfield is saturated with groundwater, it won't be able to perform its cleansing action. It's possible to improve things with better drainage. *(See p. 53 for "the French drain," also called curtain or intercept drain.)*

Two Degrees of Drainfield Failure

1. *Partial failure:* Here the drainfield will work during dry weather, but with heavy rains (or high household use), the system is overloaded.
2. *Complete failure:* Worst case scenario. Here no remedial steps can be taken with the present drainfield.

Remedies for Partially Failed Drainfields

1. Cleaning the distribution pipe in the trench may work if things have not progressed too far. Have it snaked out with a Roto-Rooter.
2. If caused by roots, dig up a section and look. Remove invasive trees and plants. Roto-Rooter pipes. Consider removing invasive trees or plants. The drainfield may recover as dead roots decay.
3. Cut water usage in half, if possible.
4. Improve drainage.

The Distribution Box Solution

Engineer Rick Duncan recommends the following procedure for a saturated drainfield (here professional help is recommended): If there are several drainfield laterals, inspect them to see if they are all being overloaded. If not and if possible, put in a distribution box with flow levellers (or weir inserts). Adjust the levellers so effluent flows mostly to under-utilized lines. The idea is to allow the over-utilized lines to rest and rejuvenate —this may take a number of months, depending on local soil and climate.

If *all* drainfields are saturated, use a larger distribution box, with inserts, and install several replacement lines, leaving the existing lines hooked up as well. Here the old lines are not abandoned and they may be useful after being rested a while.

Note: Concrete distribution boxes tend to deteriorate rather rapidly, especially in acidic soils. High-density, polyethylene D-boxes solve this problem.

Remedies for Completely Failed Drainfields

1. Hire a professional to install a new drainfield, or better yet, dual drainfields *(see p. 20)* or a pressure-dosed shallow drainfield *(see p. 22)*.
2. Do it yourself if you know what you're doing. Here you'll act as the contractor—hiring a backhoe, dumptruck for gravel, etc. However, it's very important that the drainfield be properly designed and installed—this isn't an area for guesswork. Consult local professionals.

"Repair-As-Is" Option

If you approach your local health department regarding a drainfield failure, they will often require a much larger drainfield than was originally installed, as well as a larger tank, or, in some cases, an entirely different system (a mound, sand filter, etc., along with the machinery required to run these units). There is, however, an exception, and that is the "repair-as-is" option. This is where the health department will allow you to:

- pump out the tank and ensure that the tank and all connections are functioning properly.
- install a replacement drainfield the same size or larger than the original, or a dual drainfield, or a pressure-dosed shallow drainfield. *(See p. 22.)*

This seems to us a desirable alternative, especially if, for instance, the garbage disposal is disconnected and the home residents follow practical daily domestic wastewater practices, as shown in Chapter 4 *(pp. 38–45)*. It saves money and materials, is kinder to the environment, conserves energy (gravity power vs. electricity), and is lower maintenance.

Old-Fashioned Muscle Power

We have heard reports of a small town in Southern California where a number of septic systems have been failing in recent years, and the county health department has been requiring high-tech fixes—mounds, sand filters, pumps, etc. —often costing $20,000–$30,000.

A local enterprise has sprung up: a worker has been hand-digging new drainfields, installing pipe, new gravel, and permeable membrane, for a total of around $700. At the same time, the tank is pumped, tees checked and (hopefully) the householders follow the advice in Chapter 4 *(pp. 38–45)* and end up with a few more decades of a functioning system.

Terralift Soil Loosener

This is a machine that uses pneumatic pressure to make fissures in the soil around the distribution trenches and inject polystyrene beads to promote the flow of water into the vadose (the area of soil where air penetrates). This method can be effective in certain types of soil, but check with your licensed sanitarian, contractor, or local regulatory authority. *(See p. 173.)*

CONSULTING REGULATORY AGENCIES AND ENGINEERS

The above brings up what is often a homeowner's dilemma—a "rock and a hard place" situation. What if a regulatory agency is requiring an upgrade that you cannot afford, yet you are uncomfortable with sidestepping regulations or making the repairs on your own?

There are sound historical reasons for health regulations. A century ago, when much of America was sparsely populated, relatively primitive systems, such as outhouses and cesspools, were widely used. If an unsanitary situation developed, it was confined to the farm or homestead. But as the population grew, as people began living closer together, and as health problems surfaced, regulations were enacted. For good reason.

Worst-Case Agency Scenario

The trouble these days is not with sound health codes, but with what we all may have experienced in one field or another: regulatory overkill. The worst-case scenario is an agency that will not consider anything other than a high-tech septic system. It will not allow the "repair-as-is" option, it will not countenance the standard gravity-fed septic system, nor is it open to new, lower-cost solutions such as the recirculating gravel filter, or shallow pressure-dosed drainfields. The agency may even have a list of engineers that a homeowner must choose from.

Worst-Case Homeowner's Scenario

The worst-case scenario in owner-installed or owner-fixed systems can be substandard, useless, or even dangerous repairs that can lead to civil or even criminal (!) action. Septic system design is not rocket science, yet neither is it simple or straightforward on many sites. Many systems fail because they are improperly designed and/or installed. Fixing a failed drainfield may be either simple and functional, or, if improperly done, a waste of time and materials. Another factor, of special interest if you are considering selling your home, has surfaced *(sic)* in recent years: Sellers have been sued for non-disclosure after the sale of a home and subsequent septic system failure. Seller beware!

Best-Case Scenarios

What's a poor homeowner to do? The homeowner may want to step out of the agency/engineer/contractor loop and participate in some or all of the functions of fixing an ailing system. It would be ideal if you were able to consult with the regulatory agency, which would be, in turn, open to owner-controlled repairs, subject to inspection. Often, officials will help homeowners with guidance and advice. Additionally, the services of a professional septic systems engineer can be invaluable. Although a standard septic tank and its connections are not hard to understand, its placement, the hydrological capacity of the soil, and the type and placement of the drainfield are factors where professional help is often the difference between success and failure.

We can't tell you what to do, but we hope we can make you aware of options and the range of consequences. Very often, officials will be open to a positive, well-researched owner approach. An engineer may be willing to work with you to solve your problems within your budget, telling you what functions you may perform yourself.

Note: *The National Small Flows Clearinghouse has a very good reference for failed drainfields, "Drainfield Rehabilitation."*

Go to http://www.nesc.wvu.edu/nsfc/pdf/pipline/PL_winter05.pdf

7 Graywater Systems

CHAPTER SEVEN

Graywater Systems

FIRST, SOME DEFINITIONS

Blackwater is wastewater from the toilet or kitchen sink.*

Graywater is simply everything else.

Graywater system is a system which separates graywater from blackwater to divert it away from the septic tank.

WHY DIVERT GRAYWATER?

Separate treatment of graywater can greatly reduce the load on a septic tank. It can also provide water (and nutrients in some cases) to plants; this is especially useful in times of drought or where water is in short supply (or expensive).

Washing Machine and Bath/Shower

The two largest producers of wastewater are the washing machine and the bath/shower. Laundry water is relatively easy to recycle since no changes are needed in the house plumbing. It is also relatively easy to install a diverter valve for sinks and bath/shower, so that you can toggle between a graywater system and the septic tank, as conditions warrant.

Wastewater from the toilet *and* the kitchen sink are blackwater and should go into the septic tank. Many people have tried diverting kitchen sink water into graywater systems, usually with unfortunate results (smell, clogging, frequent maintenance).

*Kitchen sink water is generally not considered blackwater, but because of the many raw food particles it contains and the problems they can cause, we put it in the blackwater category.

WHAT CAN GO WRONG: HEALTH CONSIDERATIONS

Graywater may contain infectious organisms, so keep this in mind when designing and using a system. Even though graywater is the water you just bathed in, or residue from clothes you wore not long ago, it may contain harmful microorganisms. (Bath/shower water, even laundry water, for example, may contain traces of feces.)

Safety Guidelines

All graywater safety guidelines stem from these two principles:

1. Graywater must pass slowly through healthy topsoil for natural purification to occur.
2. Graywater systems should be designed so that no contact takes place before purification.

When graywater is used for irrigating plants, it is purified naturally by biological activity in topsoil, as in a septic tank drainfield. The treatment level is probably higher, however, since biological activity is highest in the top few inches of soil. The water and nutrients are certainly more available to plants. Soil microorganisms break down organic contaminants (including bacteria, viruses, and biocompatible cleaners) into water-soluble plant nutrients. Plant roots take up these nutrients and much of the water. The pure water left over percolates down and recharges the aquifer.

IS IT LEGAL?

Legality of graywater systems varies from one location to another. In 1993, a graywater ordinance (Appendix W of the Uniform Building Code) became available as a model ordinance in 22 Western states of the U.S. In 1994, Appendix J took effect in California. An unfortunate requirement of these ordinances is that the required minimum depth of mini-drainfields is 17 to 18 inches, putting wastewater too deep in the soil for the microorganisms (in the top foot of soil) to perform their cleansing action.

HOW TO DIVERT GRAYWATER

There are a great variety of graywater systems, some of them commercially available. In the following section we present the simplest ones, most of which are geared to easy installation by a homeowner who is handy at building things. None of these utilize a pump or drip irrigation.

DISHPAN DUMP

The simplest (and most primitive) of all graywater systems is the dishpan dump. (This assumes you are using a rubber or plastic dishpan to wash dishes.) When the dishpan water gets dirty, carry it outside and dump it in a flowerbed or at the base of a tree. Simple, but hardly elegant!

GARDEN HOSE THROUGH THE BATHROOM

Another primitive solution is to block the bath or shower drain with a flat rubber drain stopper, and use warmup water to start a siphon with a garden hose. You then siphon the water out of the bottom of the tub or shower stall into the garden after you bathe or as you shower. No construction necessary. (A flow restrictor may help in this operation.)

Create an Oasis with Greywater

Art Ludwig, who helped us put together this chapter, has recently come out with his updated *Create an Oasis with Greywater—Choosing, Building and Using Greywater Systems—Includes Branched Drains.*

www.oasisdesign.net/greywater/createanoasis/

Also, see http://greywater.net/ for comprehensive information on graywater practices, common graywater mistakes, and a question and answer section.

Graywater from washing machine
Washing machine pumps water into plastic barrel (surge tank), from which it flows via gravity to mulch basin. Hose can be moved from one mulch basin to another. It is important for the water to flow immediately under mulch here, and not pond on the surface.

SURGE TANK FOR WASHING MACHINE

This is a simple and reliable way of keeping a fair quantity of water and lint out of your septic system and at the same time delivering that same amount of water to your garden. It does require that someone move the hose around in the garden, but in exchange for this one ongoing task, you have a reliable way of delivering laundry graywater to your berries, roses, shrubs, lawn, or trees.

Mulch basins are the key to efficiency and cleanliness here. They should be 6″–12″ deep, and the mulch can be mounded up an additional 6″ above grade. The mulch should be topped off annually; wood chips from tree trimmers work best.

Note: Do not route water from diaper washing into a graywater system. For obvious reasons, it should go to the septic tank.

Typically, you place a 55-gallon drum (plastic works best) outside the wall from the washing machine, possibly up on blocks to get better gravity flow to the garden. Cover the top with a lid or screen to keep mosquitoes and leaves out. Feed the washing machine hose through a window, the dryer vent, or a small hole drilled in the wall to the drum. It's best to make a basket of ¼-inch mesh at the top of the drum to catch lint. The pump from the washing machine outlet drives the water into the drum.

The drum temporarily holds water that surges out too quickly for the hose, and also allows it to cool. A ¾-inch garden hose (no smaller) is hooked up to the bottom of the tank. Water should not be *stored* in the tank, however, as it will start to smell. Every few years or so you will have to clean out a layer of smelly anaerobic sludge that has accumulated in the bottom of the tank. In winter the hose from the washing machine can be put back into the sink, and drain to the septic tank, if desired.

GRAYWATER DIVERTER VALVE

Here you install a diverter valve that allows you to send the water from the bathroom basin and/or bath/shower (*not* the toilet) into the graywater system *or* into the septic system. You have a choice. If possible, you will want to design this so you can reach this valve (or a handle extension) without having to crawl under the house.

GRAYWATER CAN BE LOADED WITH BACTERIA. IT IS VERY IMPORTANT TO GET IT *UNDERGROUND*, AND NOT HAVE IT PONDING ON THE SURFACE.

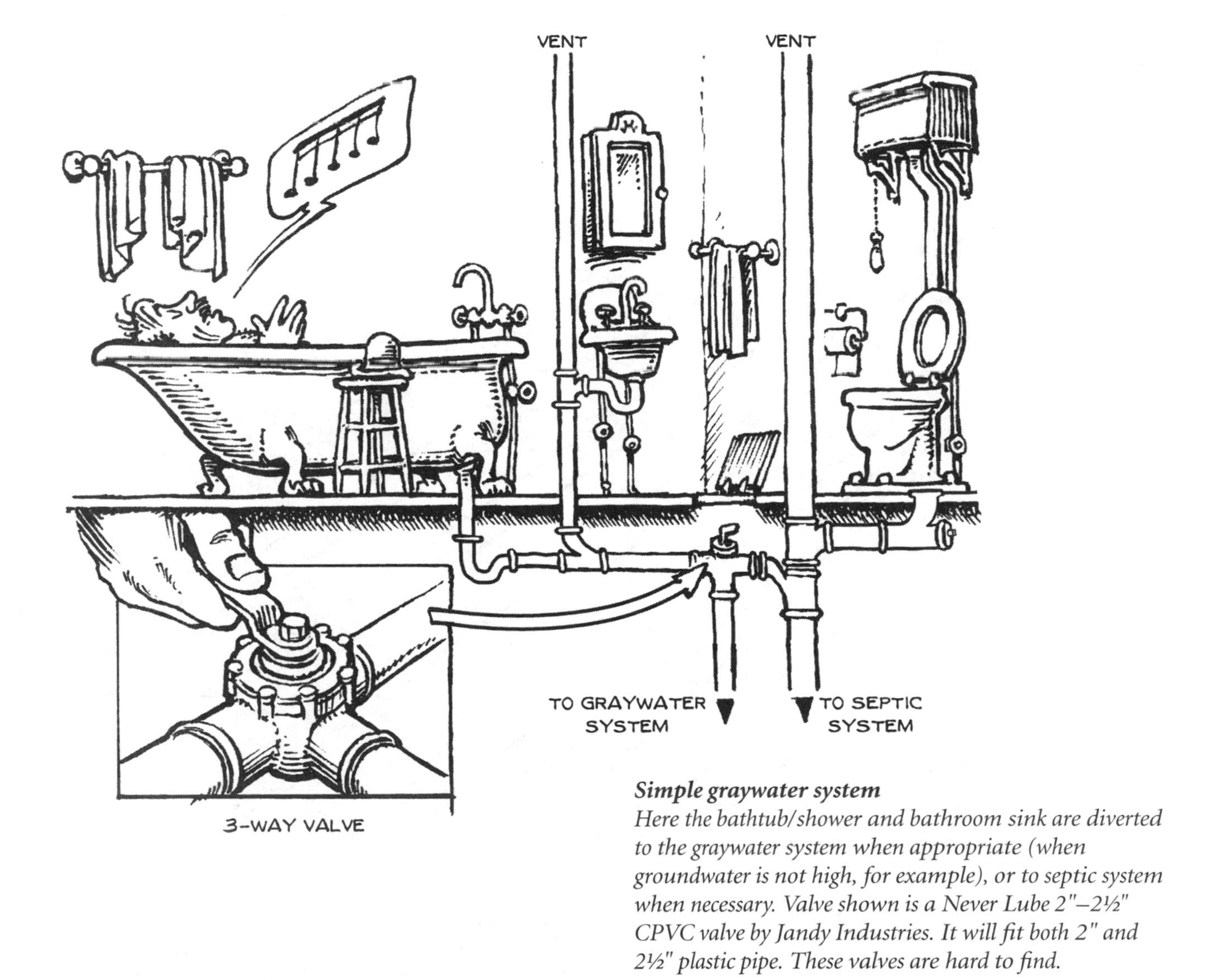

Simple graywater system

Here the bathtub/shower and bathroom sink are diverted to the graywater system when appropriate (when groundwater is not high, for example), or to septic system when necessary. Valve shown is a Never Lube 2"–2½" CPVC valve by Jandy Industries. It will fit both 2" and 2½" plastic pipe. These valves are hard to find.

EBBE'S SMALL DRAINFIELD

This system was designed by Ebbe Borregaard for washing machine discharge. It consists of a shallow underground system, 7 feet × 22 feet × 15 inches deep, with no holding tank or filtering.

Construction Details

The bottom is loosened with a digging fork (or rototiller), then layered with about 4 inches of gravel. Next a system of manifolds (feeder pipes) and drain lines of 4-inch flexible plastic perforated pipe is assembled. Pipe and gravel must be level.

The gravel is covered with landscape filter cloth to reduce silt infiltration, and the topsoil is replaced. The pipes in this system hold about 65 gallons, enough to accept almost two laundry loads of water—even if percolation is slow.

In areas of high water tables and winter saturation, this drainfield could conceivably be a mini-mound (partially above grade).

The same type of drainfield has been used by some homeowners for outdoor showers.

Note: It is important to keep lint out of this system, to avoid clogging. *See p. 175 for info on the Septic Protector.*

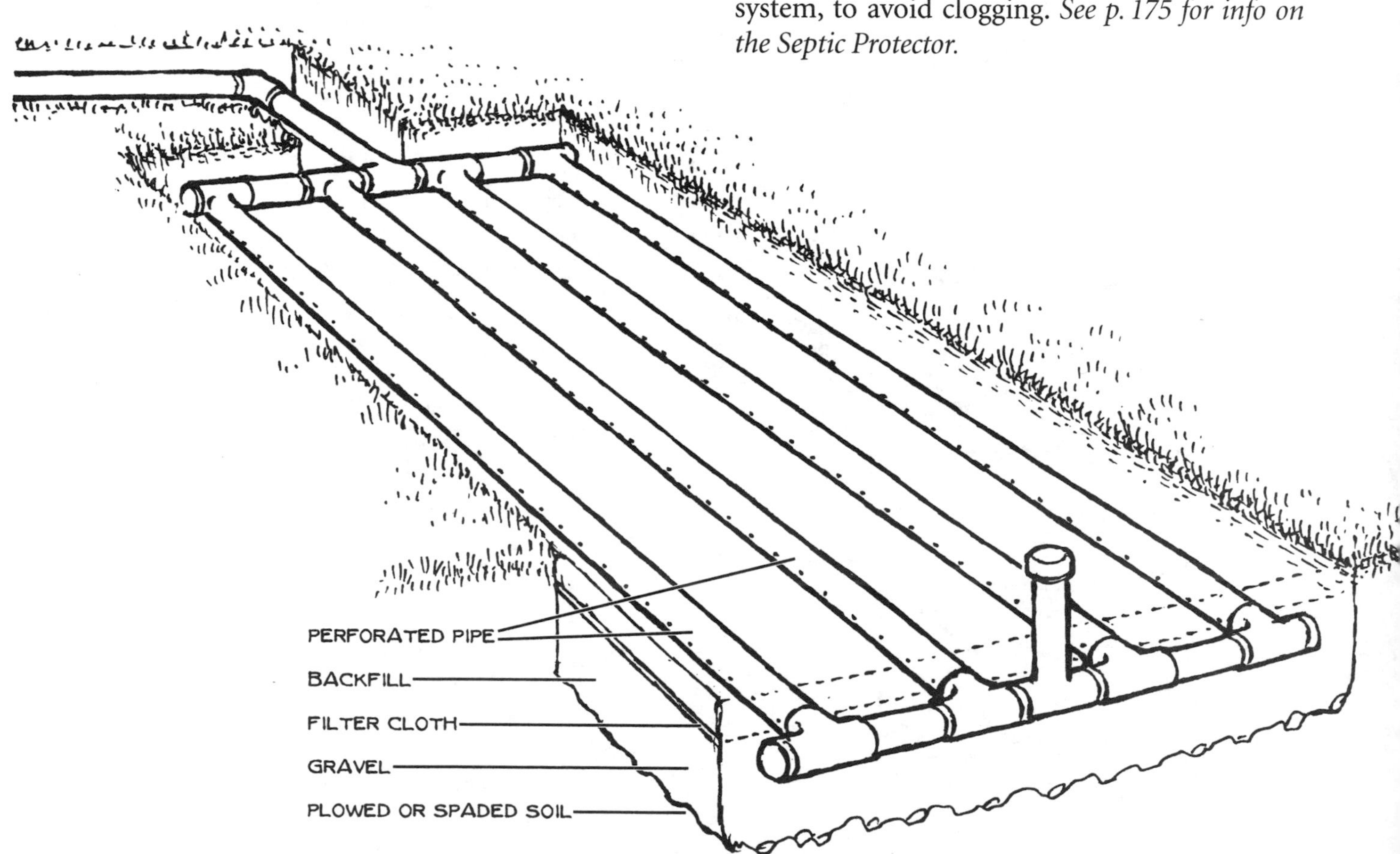

Small drainfield
This simple drainfield is set up to receive wastewater directly from shower, bathtub, bathroom sink, or from surge tank connected to washing machine.

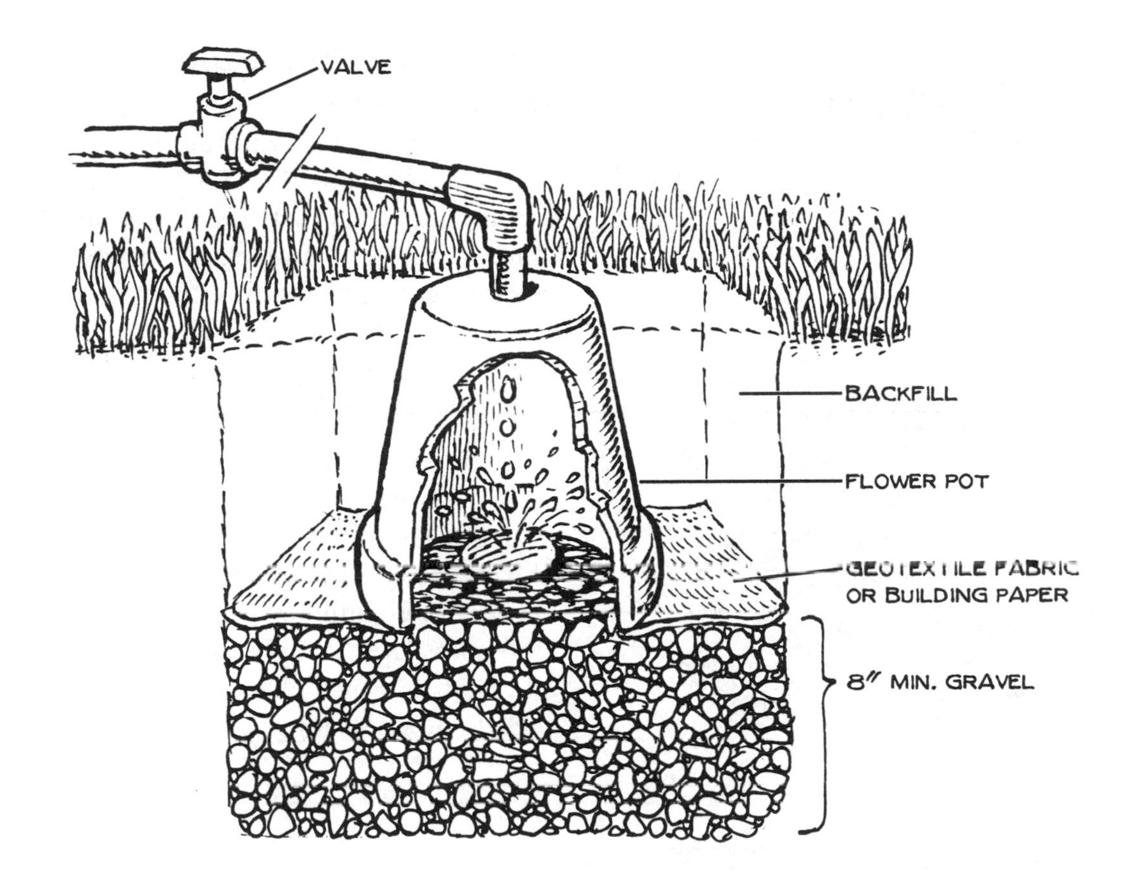

LARRY'S MINI-DRAINFIELD

Another type of small drainfield is the system designed by Larry Farwell, the man largely responsible for the legalization of graywater in the State of California. This design uses an inverted flower pot in a small drainfield. The pot top being above grade means you can move the hose from one spot to another.

Mini-drainfields are supposed to be completely buried, with gravel directly under the emitter. If you are concerned about checking for clogs and/or don't want to put gravel in the garden, mini-drainfields can be constructed to barely protrude from the surface of a hole filled with wood chips (instead of gravel). This makes service easy.

A number of these can be placed in different parts of the garden, fed by a surge tank *(see p. 66)*, and flow can be controlled by ball valves.

TWO IMPORTANT GRAYWATER PRINCIPLES FROM ART LUDWIG

- DESIGN YOUR GRAYWATER SYSTEM SO NO HUMAN OR ANIMAL CONTACT TAKES PLACE BEFORE PURIFICATION.
- GRAYWATER MUST PASS SLOWLY THROUGH HEALTHY TOPSOIL FOR NATURAL PURIFICATION TO OCCUR.

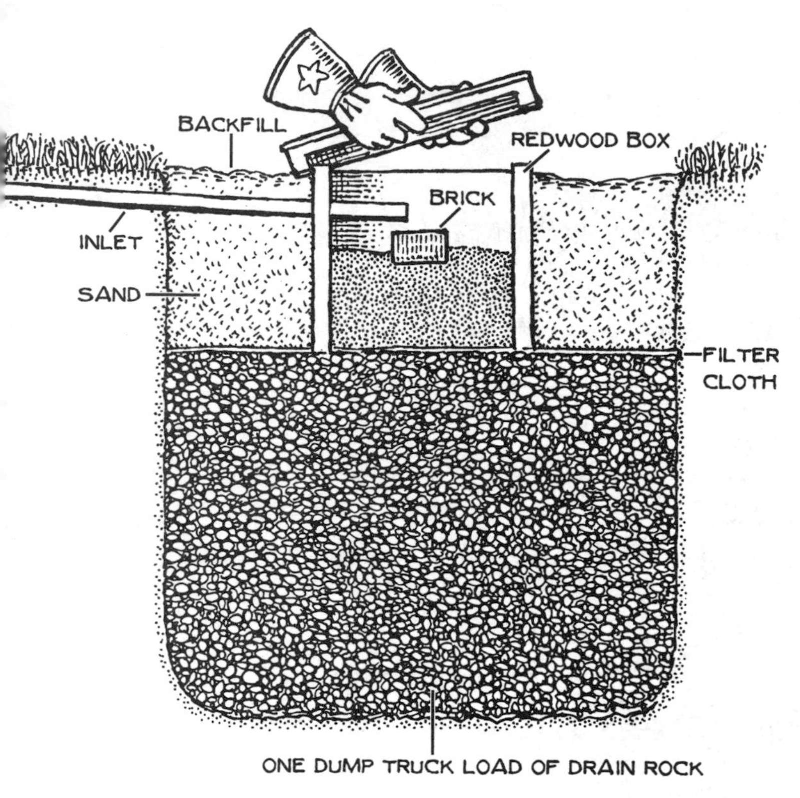

THE DUMP TRUCK DRAINFIELD

This consists of a hole 6×6×6 feet filled with one dump truck load of drain rock. On top is a redwood filter box 3×3×3 feet. Between the box and the drain rock is a 6- to 8-inch layer of sand. When the sand clogs (maybe once or twice a year) and the box starts spilling over, you scrape off the top inch or two of sand. (*Note:* never stir the sand.) After you do this several times, you'll have to put in new sand.

This system could be smaller where space is limited. Also, if the water table is high, you would make the drainfield wider and shallower. This system is meant mainly for a washing machine. The unique feature here is the use of sand as a filter to keep soap scum, lint, and other dirt out of the leaching area, so it is less likely to become clogged.

MICHAEL'S MINI–SEPTIC SYSTEM

This system, designed by Michael Gaspers, is a miniature septic system: a small, built-in-place concrete tank and a shallow drainfield. Entirely below ground, it needs very little tending.

The tank measures about 4 feet × 6 feet × 24 inches deep. It was formed and built in place. Partitions keep sludge and scum from flowing into the drainfield. One such system handled shower and laundry water for about 10 years with no maintenance. An inspection of the tank after 10 years' use revealed floating soap (retained behind the first baffle) and about 4 inches of sludge in the first compartment. There was very little sludge near the outlet. The tank was cleaned and remains in service.

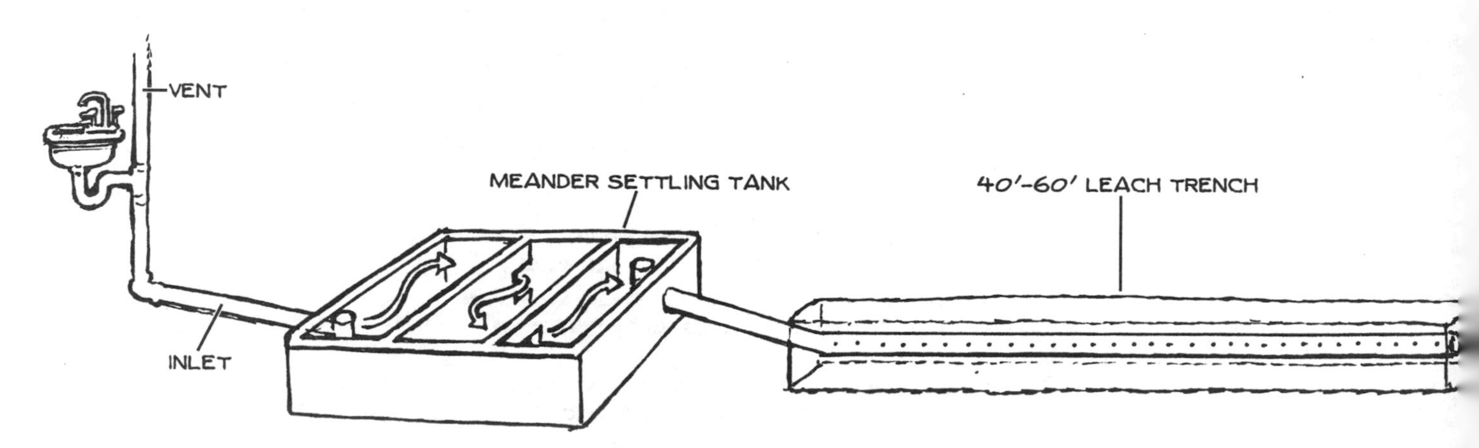

LEACH LINES UNDER RAISED BED

This system can work in tandem with raised beds, as utilized by Bio-Dynamic gardeners and popularized by gardening guru Allen Chadwick in the '70s. *Please note:* This should not be used in beds with carrots, potatoes, or other root crops. It's safest with flowers or vegetables that grow with the edible section well away from the soil.

Construction Details

4-inch drainfield pipe is buried with a capped elbow protruding to the surface. *Note: it's important that the pipe be very level.* The cap is removed and a graywater hose inserted. It is important to keep lint out of this system, to avoid clogging. *(See p. 175 for info on the Septic Protector.)*

Note: If there is a lot of clay in the soil, there must be gravel under the pipes.

If water is not filtered, the drain pipe may eventually clog, but a 4-inch pipe takes a long time to clog, and even then, it's not so hard to dig it up for cleaning (double-digging—about two shovel-heads deep—is part of Chadwick's system anyway).

Wastewater delivered to root zone
Hose from surge tank is moved around to different inlets. Caps on risers are unglued. Do not plant root crops near this system.

8 Composting Toilet Systems

CHAPTER EIGHT

Composting Toilet Systems

by Carol Steinfeld

This garbage heaped up beside the stone blocks, the tumbrels of mire jolting through the streets at night, the awful scavengers' carts, the fetid stream of subterranean slime that the pavement hides from you, do you know what all this is? It is the flowering meadow, it is the green grass, it is marjoram and thyme and sage . . . it is perfumed hay, it is golden wheat, it is bread on your table, it is joy, it is life.

–Victor Hugo, Les Miserables, *1862*

When a full septic system is not feasible, such as in remote areas or where there is poor soil drainage, an alternative system can include a composting toilet system.

Composting toilets (also known as dry, biological, or waterless toilets) use the same biological process as that at work in a yard waste composter to oxidize and break down blackwater (excrement and toilet water) into a stable form that looks like soil, not sewage.

Once considered an option only for parks, back-to-the-landers, and seasonal cottages, composting toilet systems are now turning up in mainstream homes and buildings. Increasing their acceptance is the adoption of microflush toilets and graywater-only systems for the rest of the wastewater equation. Also, many owners now opt for service contracts to maintain the systems.

Composting relies on air-using, or *aerobic,* bacteria, which work 10 to 20 times faster than the anaerobic bacteria at work in septic tanks. So, the challenge of composting toilets is getting air to the composting process while minimizing human exposure to it. That requires careful engineering of air flow so that air is taken in and then exhausted through the exhaust chimney, not the toilet room. It also requires some management of the composting material: either turning or batching the material and adding chunky carbon material to keep it porous, so the aerobic bacteria are healthy and functional. Keeping the material aerated also means that it can't be saturated, so only waterless or microflush toilets are used with composting toilets.

Composting toilets first enjoyed relatively wide play in the United States in the 1970s in response to new awareness of water pollution caused by wastewater. A few companies and nonprofit organizations introduced manufactured models and build-it-yourself designs. However, the need for testing became evident when several systems proved faulty, due to shoddy construction or poor designs that revealed a misunderstanding of the requirements of composting. There were reports of odors, flies, incomplete processing, hard-to-empty systems, and compacted organic material. Formal damnation came in the early 1980s from the U.S. Environmental Protection Agency and the California Department of Health in the form of a report that identified flaws resulting from poor design, installation, and construction. After that, composting toilets were mostly relegated to remote country sites. Chalk up this era to "lessons learned." Since then, most systems have improved significantly. Better site-built designs are now available and reflect a healthier respect for the aeration and heat requirements of composting.

Today, the availability of complementary microflush toilets, graywater (washwater) systems, and maintenance contracts have improved their acceptance, and have made these systems ready for prime time.

WHY INSTALL A COMPOSTING TOILET?

Essentially, composting toilets are used to divert excrement from the wastewater mix. In many states, this means that installing a composting toilet system allows a property owner to install a graywater-only drainfield 35 to 40 percent smaller than a full-sized one, saving money and preserving landscapes.

Other reasons property owners usually install composting toilets include:

- Poor soil drainage or geological conditions do not allow a septic system.
- High groundwater or proximity to streams and shorelines limits what they can install.
- Water is expensive or unavailable for flushing toilets.
- An undersized or failing septic system needs to be supplemented.
- Ecological awareness prompts the desire for a non-polluting system.

In their current forms, composting toilet systems are not a flush-and-forget technology. They require a consciousness of what's put into the toilet (no toxic chemicals, please!), some maintenance, well-thought-out siting and installation, and, usually, extra electricity for operating fans and heaters.

Some people love them, and some hate them—usually based on their experiences with a particular system. Some systems are just poorly installed. Perhaps the most common installation mistakes are siting them in cold places (unheated basements, for example), not draining away extra moisture (called *leachate**), and installing systems that are too small for the usage they'll get.

Composting toilet systems range from little 7 gallon composters for boats to large aerated two-bin systems that serve more than 100 people daily. They are most prevalent in coastal cottages in Scandinavia, as well as in Vietnam and Mexico. (There are more than 100,000 twin-bin-style composting toilet systems in Vietnam and Mexico today.)

The technology varies widely, but the typical components of a composting toilet system are:

- a waterless toilet stool or a microflush toilet
- a composter to which one or more dry or microflush toilets flow
- a screened air inlet to provide air for the process
- an exhaust system, often fan-forced, to remove odors, carbon dioxide, and water vapor
- an access door to remove the end-product
- (optional) process controls, such as mixers
- (optional but recommended) a means of draining and managing excess leachate

*Leachate is the extra liquid—usually urine and flush water—that drains to the bottom. Some confuse it with "compost tea," which is an oxidized liquid from plant waste composters. Leachate is much stronger stuff, and can contain high nitrogen, acid, and fecal coliform. It can kill your plants.

FLUSH TOILETS AND DRY TOILETS TO USE WITH A COMPOSTING TOILET SYSTEM

To avoid the "black hole" that one encounters when looking down a dry toilet that flows to a composting reactor, more and more buyers of the larger composting systems are installing them with microflush toilets. Originally designed for boats, these use one pint of water per flush. Other innovative toilets include the vacuum toilets and a toilet that uses one teaspoon of water simply to lubricate a rotating cup that revolves to drop the deposited excrement out of sight and into the composting reactor. With microflush toilets, the flush water produces extra leachate that must be managed. (Methods for this are described later in this chapter.) Also, flush water can cool the contents of the composter, so extra heat is needed to maintain faster composting rates.

SELF-CONTAINED VS. CENTRAL COMPOSTING TOILETS

Composting toilet systems are either *self-contained* (usually for cottages) or *central* (also referred to as *remote* and *below-floor*). With self-contained systems, the toilet seat and a small composting reactor are all one unit. These can sit in the bathroom. Due to their small size, they are typically used in cottages and seasonal homes. Their capacity typically ranges from two to six adults, varying with the model. Prices range from about $750 to $1,500.

In a central system, the toilet flows to a composting reactor that is somewhere else, usually located in the basement or in its own enclosure to the side of the building. Commonly the choice for year-round homes and facilities with multiple toilets, they range in price from $1,400 to $10,000 and higher for institutional-capacity systems.

Self-contained composting toilets were developed for and are mostly purchased by cottage owners. Remember that in a self-contained system you can see the contents up close and personal. This usually isn't terribly offensive, thanks to the additive you put in it, but it may be off-putting to your mother-in-law. Too often, people buy self-contained composting systems for year-round homes. Then they discover that a small composter means more management than anticipated. Self-contained systems are usually not appropriate for year-round use for more than two people.

BATCH VS. CONTINUOUS PROCESSING

Every composting system uses either a *single-chamber continuous composting* or *multi-chamber batch composting* process. A continuous composter features a single chamber, into which excrement is continually deposited at the top, and from which finished compost is removed from the other end of the unit. The Phoenix is a continuous composter with rotating tines (turned with a hand lever) and a unique system of air circulation; the Clivus Multrum, a well-known brand with a sloped-bottom design, is another continuous composter. Some Sun-Mars and BioLets are also continuous composters.

Batch composters (such as EcoTech Carousel, Vera, BioLet NE, Wheelibatch, and two-vault systems) utilize two or more interchangeable composting reactors; one is filled, then allowed to completely compost while another reactor fills. The advantage of batch composting is that advanced compost is not contaminated by fresh waste. Also, in some systems, the composter containers can be removed from the toilet to take outside to empty.

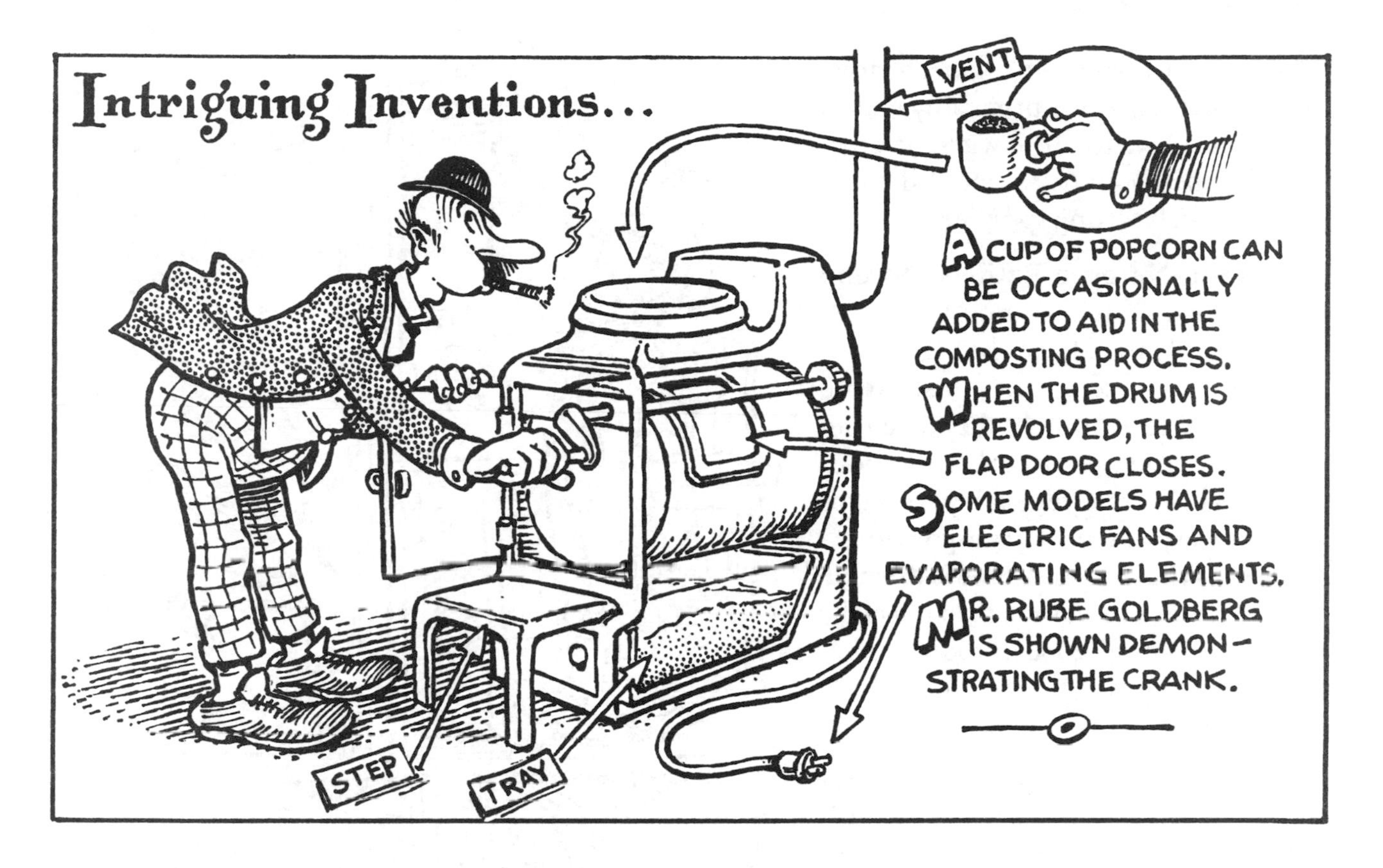

The Build-It-Yourself Option

One can either purchase a manufactured composting toilet system or build one on-site. It can be difficult to get health department approval for site-builts, however, as little performance data exists for these.

The three most common types of site-built designs are twin-bin (or two-vault) systems, alternating drum systems, and the inclined vault system.

MAINTENANCE

As mentioned earlier, composting toilet systems require periodic maintenance (some more than others), although service contracts are available from some manufacturers and septage haulers. Generally, the maintenance considerations are:

Heat: Composting is most efficient at temperatures of at least 65° F—the higher, the faster, up to 135° F. It used to be thought that the composting process would generate a lot of heat—alas, not so. Too often, composters are installed in cold basements. Good places for installation are passive solar-heated areas, next to appliances that give off waste heat, or heated rooms.

Aeration, Mixing, and Additives: Composting microorganisms need air. Airflow is critical (there should not be too many bends in the ventilation pipe). Aeration can be improved by mixing the material, adding wood shavings or popcorn to it to create air spaces, and by batching it.

Microorganisms: Composting is carried out by bacteria and fungi. These microorganisms thrive best in a warm, moist environment (but not immersed in water!). The composter can be inoculated with purchased microbes or a shovelful of finished compost.

Moisture: Microorganisms need moisture to do their work, preferably within the range of 40 to 70%, with the optimum being about 60% (about the texture of a well-wrung sponge).

Carbon-to-Nitrogen Ratio (C:N) and Additives: For the composting microbes to fully transform the high nitrogen content of excrement (mostly from the urine) to compost, they need an adequate amount of carbon—about 30 parts of carbon for each part nitrogen. However, as most urine goes to the bottom of the composter and is drained, this ratio is less of an issue than is usually reported. Typical good carbon sources are untreated bark mulch, wood shavings, rice hulls, oak leaves, etc.

Management: As with all wastewater treatment systems, management is critical to the efficiency of the system. You've got to monitor the level and consistency of material in the pile periodically.

Removing the End-Product: Compost should be removed periodically, anywhere from every three months for a cottage system to every two years for a large central system.

"BIOLOGICAL DECOMPOSITION IS THE PROCESS THAT KEEPS BIOLOGICAL WASTE MATERIAL FROM PILING UP IN THE WORLD. FALLEN TREES ROT AWAY AND DEAD SQUIRRELS DECOMPOSE. MOSTLY, BACTERIA AND FUNGI ARE MAINLY RESPONSIBLE FOR THIS BREAKDOWN OF ORGANIC MATTER, BUT OTHER ORGANISMS SUCH AS EARTHWORMS AND INSECTS CAN SPEED UP THE PROCESS. THE DECOMPOSITION OF HUMAN WASTE MATERIAL IN A COMPOST TOILET IS BASICALLY THE SAME PROCESS AS THE DECOMPOSITION OF LEAVES IN THE FOREST. COMPOST TOILETS ARE AEROBIC UNITS IN WHICH THE WASTES ARE DECOMPOSED TO CARBON DIOXIDE, WATER, AND INERT SOLIDS. AEROBIC CONDITIONS ARE REQUIRED TO MAINTAIN PROPER FUNCTIONING. IF THE COMPOST TOILET IS WELL MAINTAINED, THE PRODUCT OF DECOMPOSITION SHOULD SMELL LIKE FOREST SOIL."

—*ONSITE WASTEWATER DISPOSAL*, RICHARD J. PERKINS, LEWIS PUBLISHERS, CHELSEA, MI, 1990.

WHAT TO DO WITH THE END-PRODUCT

Finished humus has the consistency of composted leaves, and should smell earthy but not offensive. Most states require sending it to a treatment facility or burying it under at least 12 inches of soil, preferably within the root zones of non-edible plants that can use the nutrients. Many users place the composted material outside in a pile to finish composting, and add more mulch and yard waste.

Some composting toilet owner-operators use their compost on their edible vegetable gardens. If no one is sick who used the composting toilet, it's unlikely that pathogens will be a problem. But a better rule of thumb is: *don't use it on edibles.* There are plenty of non-edible plants in any garden that will be grateful for compost.

If any leachate is drained from composters, it must, by law, either be disposed of in a septic tank, removed by a septage hauler, or taken to a treatment plant for further treatment. Some innovative systems combine it with graywater for use in subsurface irrigation of plants.

PATHOGENS

A pathogen is a microorganism capable of producing a disease. Particularly in developing countries, the risk of spreading pathogens is a concern. There are two ways pathogens are destroyed in composting toilets:

- **Antibiosis:** Microbial and other higher-order aerobic organisms develop in the compost pile during the decomposition process, actually creating antibiotics that kill pathogens.
- **Time:** Out of their hosts and favored environments, pathogenic microorganisms will eventually die.

ODOR

The possibility of odor varies with the system. A system with a remote composter coupled with a toilet with a trap (such as a microflush toilet) rarely, if ever, produces any odors. A self-contained composting toilet is more apt to produce them. A good fan venting in the right direction should prevent this—in fact, this fan often vents odors in the entire bathroom, and there's rarely a lingering legacy of anyone's bathroom visit. However, a hard draft of wind can push exhaust down the vent and into the bathroom. If odors are a problem, turn up the fan or check out odor filters.

FLIES

Occasionally flies are a problem in some self-contained composting toilets and those with open, dry toilet stools. Some ways to manage: sprinkle diatomaceous earth on the composter contents, avoid putting kitchen scraps down the toilet, or introduce parasitic wasps. If flies persist, make a fly screen insert to fit inside the toilet opening. This can lift with the lid. Seal the toilet opening during non-use periods, but only if there is a screened air intake elsewhere.

DIVERTING URINE

Urine is full of nitrogen and usually sterile. Some studies report that urine and feces combine to produce a malodorous substance, and the two are easier to manage separately. For that reason, toilets that divert urine are increasingly used with composters. In Sweden, several urban and suburban homes and apartment buildings with composting systems feature toilets that divert urine to storage tanks to be later used in agriculture. You can consider adding urine to a subsurface, graywater irrigation system, as graywater contains a lot of carbon but not much nitrogen. But be sure to plant it with salt-loving plants (halophytes) as urine and graywater are very salty!

COSTS

The costs of manufactured composting toilets vary widely. For a year-round home of two adults and two children, a composting toilet system could range anywhere from $1,200 to $6,000. Cottage systems designed for seasonal use range from $750 to $1,500. Large-capacity systems for public facility use can be as high as $20,000 and more. Site-built systems, such as cinder-block, double-vault systems, are as expensive as their materials and construction labor costs. Other costs include electrical work and installation labor.

GETTING THEM APPROVED

One of the major obstacles to installing a composting toilet is getting permits. Composting toilets usually require special permits from town, county, or state officials, and the ease of obtaining permits varies widely across the country. Many health officers are simply unfamiliar with these systems. Some fear that zero-discharge systems will allow building in previously undevelopable areas. Some states require back-up conventional wastewater systems. States that are more amenable than others include Massachusetts, Minnesota, Washington, Maine, and New Mexico. Permits are rarely granted for owner-built composting toilets, unless they have rigorous monitoring or maintenance plans. Many states require certification by NSF International, essentially the "Underwriter's Laboratory" of the wastewater system industry.

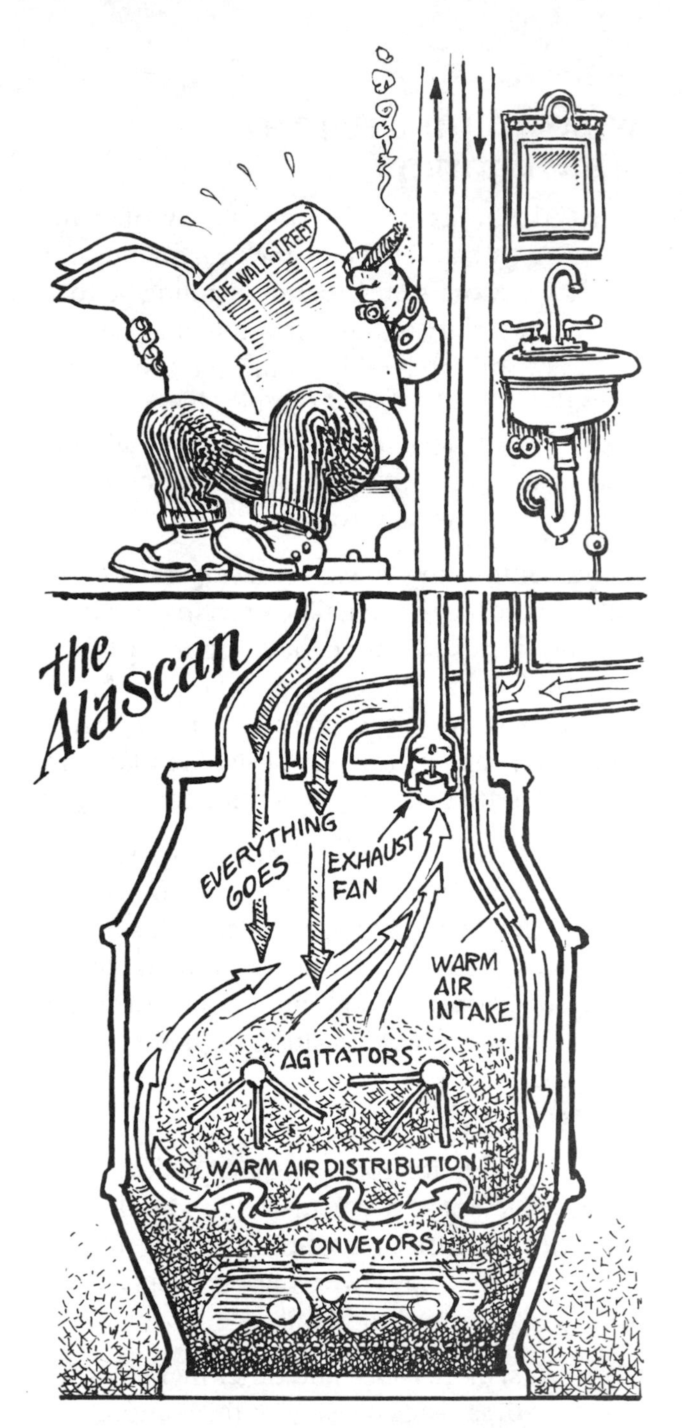

The Alascan Organic Waste Treatment System
This composting system featured two motor-driven agitators that mix in fresh waste and keep the compost mass level. Three ratchet-driven agitators at the bottom remove finished compost.

NEW COMPOSTING TOILET SYSTEM BOOK

CAROL STEINFELD, WHO WROTE THIS CHAPTER, IS CO-AUTHOR OF *THE COMPOSTING TOILET SYSTEM BOOK*. IF YOU ARE CONSIDERING A COMPOSTING TOILET, WE HIGHLY RECOMMEND GETTING THIS BOOK. IT DESCRIBES OVER 50 SYSTEMS, BOTH MANUFACTURED AND SITE-BUILT, IN GREAT DETAIL. *(SEE P. 169 FOR DETAILS.)*

DESCRIPTIONS OF SYSTEMS COMMON IN NORTH AMERICA

Among the central composters, the Phoenix is a tall polyethylene vault with a three-part system. Waste falls to a high area, where rotatable tines act as a mixing device to break it up and aerate it. It then falls to a grate, then to a collection box. The Clivus Multrum, CTS, and Clivus Minimus (site-built system) are sloped-bottom tanks in which waste moves down an incline, which ideally slows its passage to the bottom, helping to aerate it. Compost is removed through an access hatch at the bottom of the tank. The AlasCan is a highly mechanized composter with power-driven auger-agitators and high-velocity air flow via a fan. The Bio-Sun system is a large canister tank in which a powerful air compressor moves large volumes of air across the excrement. Increasingly, Bio-Sun is designing the composting reactor right into the foundations of buildings. (In some buildings, the entire basement is the composter.) Two-vault systems, built throughout the developing world, consist of two vaults, often made of concrete block, that are used interchangeably: when one side fills, the toilet stool is moved to over the other (or a valve redirects the flow from a fixed toilet).

Occasional leveling of the material with a pitchfork is required in some of these systems. Users also turn the compost at the removal hatch end to prevent compaction and promote aeration. In all of these models, leachate—urine and other liquids—drains to the bottom, where it is evaporated or must be drained for disposal or utilization. Some employ a leachate-recirculating system that pumps liquid from the bottom and sprinkles it on the top of the material.

The EcoTech Carousel is a fiberglass cylindrical container consisting of an outer container that holds an inner container divided into four compost chambers. This keeps leachate separate from the compost, and keeps compost in batches. When one chamber fills up, the next is rotated into position. The Vera Toga 2000 is a system of roll-away interchangeable 60-gallon compost reactors. Extra containers can be purchased, giving the system as much capacity as one has containers. The Wheelibatch system is a site-built version of the same, made of roll-away trash bins fitted with aerators. Some site-built systems utilize used 50-gallon drums. Sun-Mar's new Centrex Plus features a two-chamber version of its patented *bio-drum,* a rotatable canister, and a finishing drawer. Sun-Mar's Centrex, a smaller unit featuring a smaller bio-drum and just one collection tray, is typically only used in cottages.

Among the self-contained composters, Sun-Mar's series of composting toilets all feature a revolving bio-drum—a canister-like composter mounted horizontally. A hand crank allows users to periodically rotate the bio-drum to mix and aerate the material. BioLet's XL uses a mixing arm to slice through the composting waste and push it through a grate. Vera's Toga series feature interchangeable composting chambers (think: pails), as does the BioLet NE. Sancor's Envirolet has a moveable grate—called a "mulcherator"—that can be manually pulled to break up, mix, and aerate the waste.

Although the small, self-contained composters are lower-priced and have few installation problems, there are disadvantages. Users must constantly remove the finished compost while adding new material. Due to their small size, they are often overloaded and processing can be incomplete. Excess leachate is the bane of the smaller ones, and one must guard against letting the material bake and dry into a hard mass.

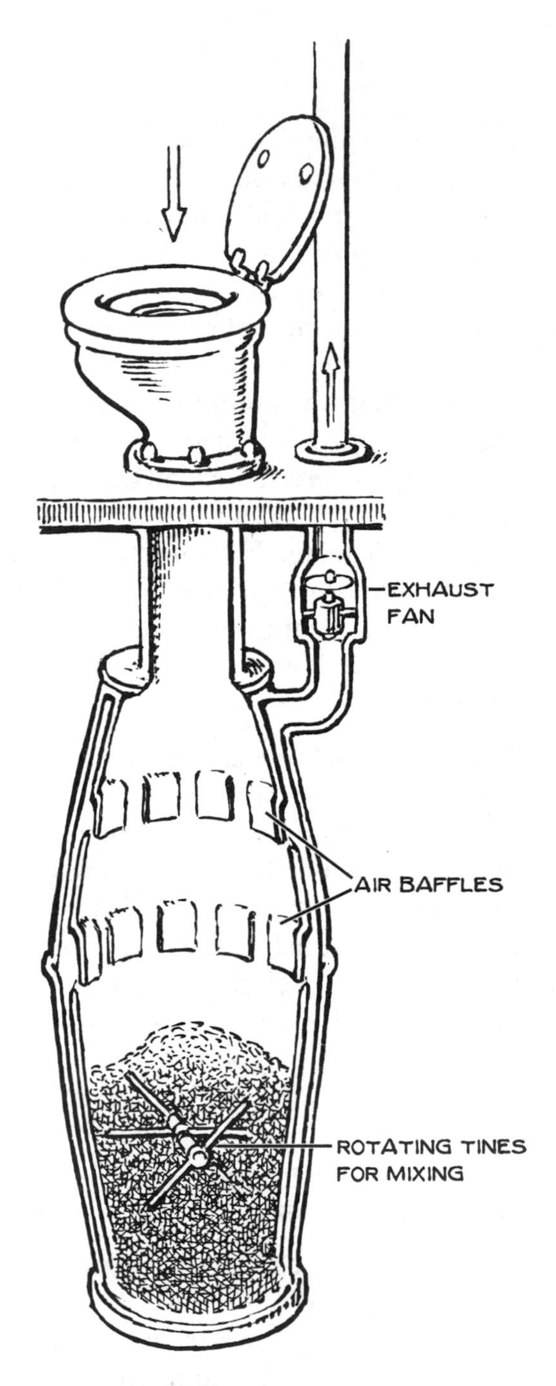

Phoenix Composting Toilet System
Bulking agent, such as wood shavings, is added for aeration and additional carbon. Tines are rotated manually to mix material. Note air baffles inside tank for adequate oxygen intake, and fan in vent for exhausting odors.

A COMPOSTING FUTURE?

Composting toilets are fast losing their reputation as rustic toilets, and due to growing awareness of water pollution sources, are getting more serious attention from regulators, public health officials, and property owners.

They're becoming more high-tech, with sensors and monitors that automatically adjust for heat, air, and moisture.

In the future, composting toilet systems may be serviced by central management districts, and the end-product taken to a central composting facility.

THE KEYS, THEN, TO SUCCESSFUL COMPOSTING TOILET SYSTEMS ARE:

1. A WELL-DESIGNED UNIT
2. PROPER INSTALLATION
3. CONSCIENTIOUS MAINTENANCE

NORTH AMERICAN COMPOSTING TOILET MANUFACTURERS AND DISTRIBUTORS

BioLet
150 East State Street
Newcomerstown, OH 43832
800-524-6538
www.biolet.com

Clivus Multrum
15 Union Street
Lawrence, MA 01840
800-425-4887
www.clivus.com

CTS Composting Toilet Systems
P.O. Box 1928
Newport, WA 99156-1928
888-786-4538
www.comtoilet.com

Vera / EcoTech Carousel
508 Boston Post Road
P.O. Box 415
Weston, MA 02493-0003
978-369-9440
eco-tech@ecological-engineering.com
www.ecological-engineering.com/ecotech.html

Envirolet Composting Toilet
SanCor Industries
140-30 Milner Avenue
Toronto, Ontario
M1S 3R3 Canada
800-387-5245
info@envirolet.com

Phoenix Composting Toilet
Advanced Composting Systems
195 Meadows Road
Whitefish, MT 59937
406-862-3854
www.compostingtoilet.com

Sun-Mar Corporation
600 Main Street
Tonawanda, NY 14150
888-341-0782
compost@sun-mar.com
www.sun-mar.com

Other Sources

The following catalogs feature two or more brands of composting toilets, as well as accessories.

Ecos Catalog: Tools for Low-Water Living
978-369-9440
eco-tech@ecological-engineering.com
Several brands of composting toilets and other water conservation tools.

Real Goods Renewables Catalog
200 Clara Street
Ukiah, CA 95482
800-919-2400
www.realgoods.com
Features all kinds of solar and off-the-grid items.

Lehman's Non-Electric Catalog
One Lehman Circle
P.O. Box 41
Kidron, OH 44636
330-857-5757
www.lehmans.com
Traditionally aimed at the Amish market, *Lehman's* offers all kinds of basic, hard-to-find tools and two or three types of composting toilets.

For More Information, Including Sources for Site-Built System Plans

For sources of plans for site-built composting toilets, see:

The Composting Toilet System Book,
by Carol Steinfeld and David Del Porto.
Published by the Ecowaters Project
P.O. Box 1330
Concord, MA 01742
978-318-7033
www.ecowaters.org

The Humanure Handbook
by Joseph C. Jenkins
Published by Jenkins Publishing (PA), 2005

9 Advanced Systems

CHAPTER NINE

Advanced Systems

The beauty of a typical standard system—comprised of a septic tank and a drainfield—is that it uses no electricity or mechanical devices. Aside from periodic pumping of accumulated septic tank solids, the system operates by natural processes. Gravity—that deceptively elegant and often overlooked principle—provides all the power needed for water and wastes to flow through the system. Treatment and disposal of the wastewater is accomplished by natural physical, chemical, and biological processes in the soil of the absorption system.

Given adequate site and soil conditions, septic systems with gravity-flow drainfields can provide successful sewage treatment and disposal for decades—some say practically indefinitely—when properly constructed and maintained.

Just what is an "advanced system"? What we call here "advanced systems" are those used where the conventional gravity-fed system will not provide adequate treatment, typically due to insufficient land, high groundwater, proximity to rivers, bays, lakes, etc., poor soil, or shallow soil over bedrock. An advanced system is, in the broadest sense, one which incorporates some modification of, or addition to, the standard gravity-powered tank and drainfield setup. A wide variety of advanced systems have been developed in response to the needs of different site conditions. Since these systems are used for sites with limited soil or other problems, the margin of safety (otherwise afforded by optimal soil and site conditions) is limited, and any failure is likely to be difficult (and expensive) to correct.

TYPES OF ADVANCED SYSTEMS

In the original version of this book, we described five advanced systems:

1. Dosed-flow systems
2. Mound systems
3. Sand filters
4. Gravel filters
5. Wetlands

The idea was to tell homeowners how each works, for two reasons:

1. So that if you were involved with construction of a new system you would understand some basic principles and be better informed when talking to regulators and engineers
2. Or in case you already had an advanced system, you would understand how to maintain it properly

In this revised edition we have amended the description of the above five systems, added a number of new (to us) advanced systems, and given references for more complete information.

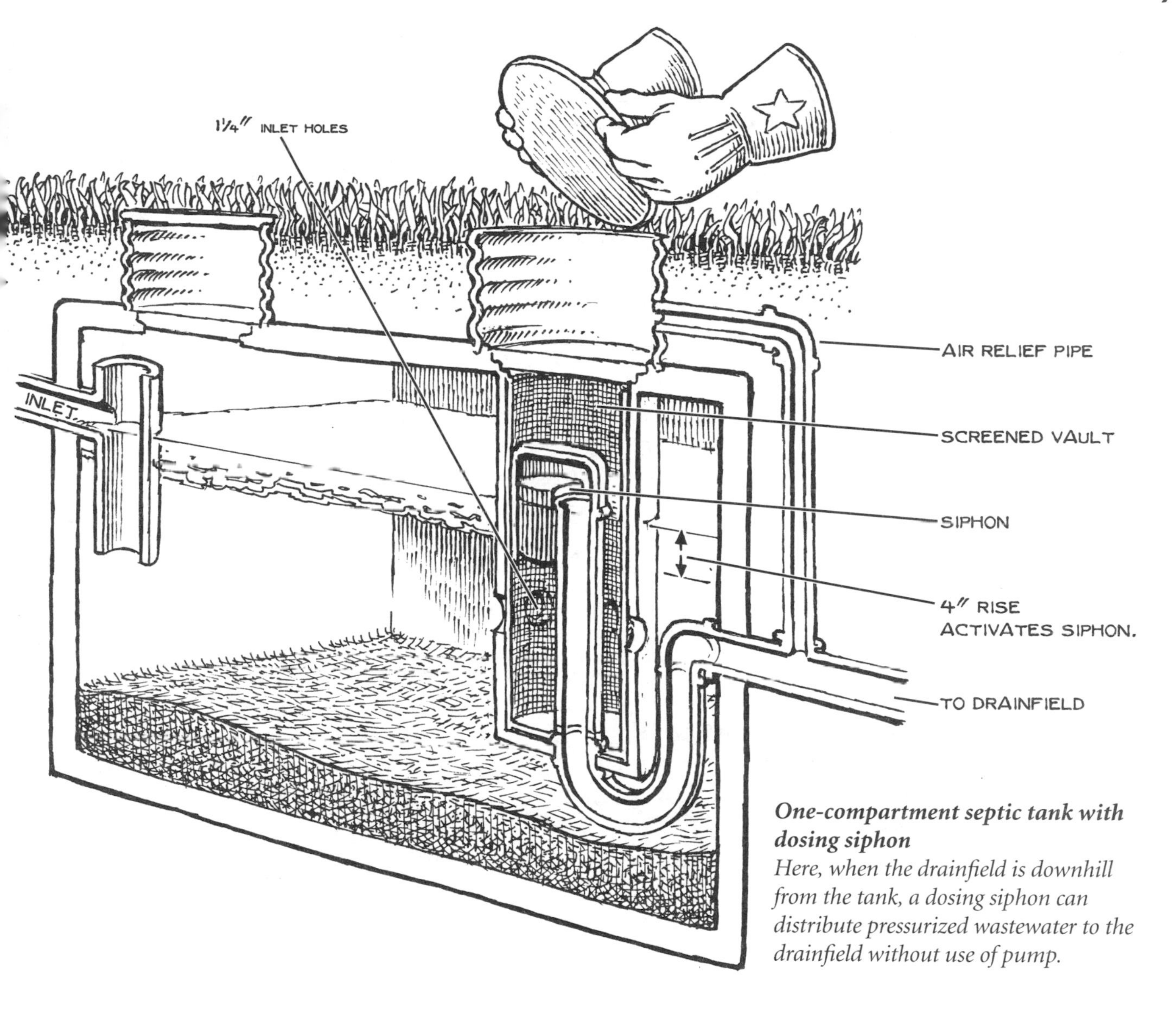

One-compartment septic tank with dosing siphon
Here, when the drainfield is downhill from the tank, a dosing siphon can distribute pressurized wastewater to the drainfield without use of pump.

DOSED-FLOW DRAINFIELDS

"Dosed-flow" refers to controlling the flow of effluent to the drainfield, as opposed to the continuous gravity flow of conventional systems. In addition to evenly distributing the water in the drainfield, timed doses allow the system to rest and recover between loadings, and avoid peak hydraulic loads on the drainfield. This helps maintain a stable biomat and prevents the oxygen in the vadose zone (area where air is available to soil) from being depleted. New systems of tanks, level sensors, controllers and pumps have made the systems more reliable than they were in the past.

Dosing Without Pressure

Here effluent is distributed to the drainfield in periodic, large-volume doses. Dosing is achieved by a pump *(see p. 88)* or, if the drainfield is located far enough downhill from the tank,

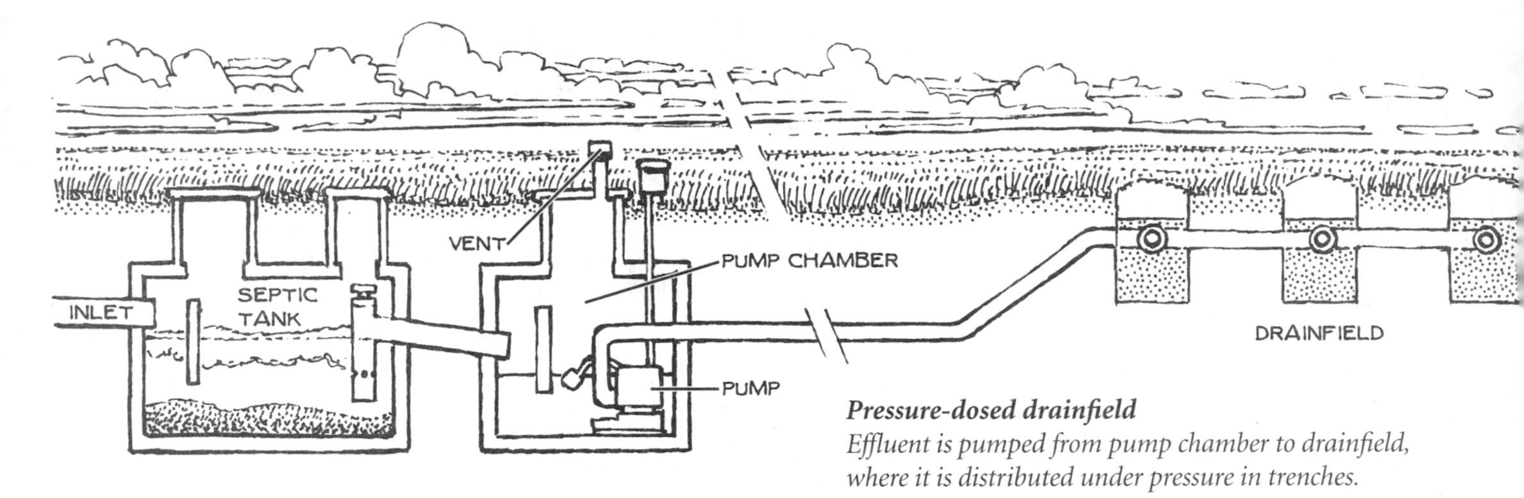

Pressure-dosed drainfield
Effluent is pumped from pump chamber to drainfield, where it is distributed under pressure in trenches.

with an energy-free dosing siphon. Effluent accumulates in the dosing tank (pump chamber) up to a predetermined volume, at which point the pump or siphon is activated, and the accumulated effluent goes to the drainfield in a single dose. The pump or siphon then shuts off until enough effluent has again accumulated in the dosing tank. There are also dosing systems with timers that dose periodically. There is also the "flood-dosed" system that is designed so that the pump fills all the pipes at once, thus utilizing the entire field. Some tanks are designed with the pump chamber in the same tank instead of using two separate tanks.

Dosing with Pressure

A pressure-dosed system, also called a pressure distribution system, provides for periodic dosing of large volumes of effluent. All of the pipes within the drainfield are filled during each dosing cycle, and a uniform volume of effluent is distributed out of each hole in the network; this allows the soil to drain between applications. Drainage brings air into the drainfield, and reduces excessive biomat growth and soil clogging. This even distribution of effluent over the entire drainfield length provides more contact with soil organisms and, therefore,

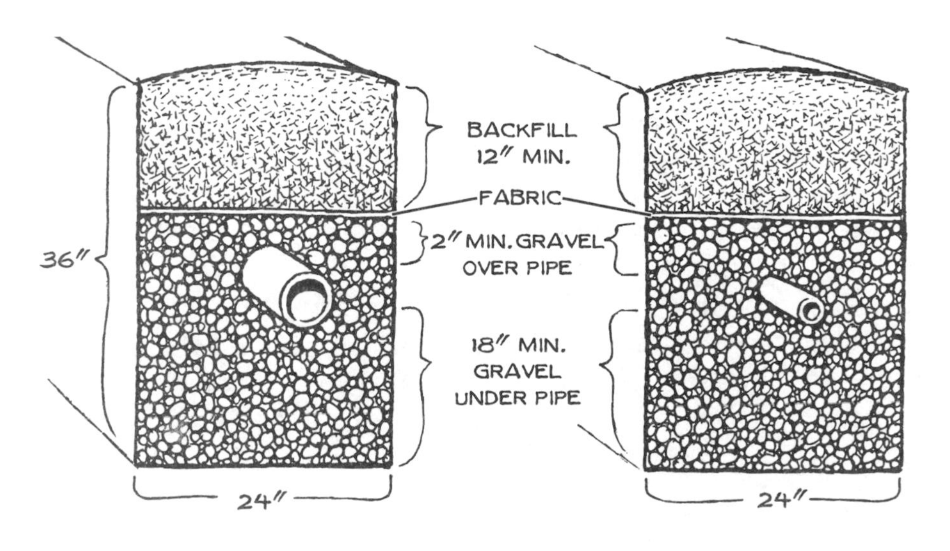

Comparison of typical gravity-flow drainfield (l) and pressure-dosed system (r)
Left: Pipe is 3"–4" diam. with one or two rows of ½"–⅝" diam. holes along bottom, spaced 3"–6" apart.

Right: Pipe is 1"–2" diam. with one row of ¼"–½" diam. holes along bottom, spaced 1"–3" apart.

enhanced treatment. This uniform distribution allows the soil to remain well aerated and prevents soil clogging from excessive loading.

MOUND SYSTEMS

Update note: In our opinion, mounds are, in most cases, no longer an appropriate technology. Many of the devices listed in this chapter provide superior treatment to mounds, at lower cost, and with much less environmental destruction. Yet, so many mounds have been built that we offer the following description for homeowners who have to maintain them.

Mounds are basically raised drainfields. They were developed in the 1940s for sites with impermeable soils, high groundwater, or other limiting conditions. They were used extensively until the last decade, when a combination of mound failures, high costs, and newer, more efficient technology, has superceded them (in areas where regulators are up-to-date). If you are in an area where mounds are still required, look through the alternatives below, and initiate discussions with your health officials.

A MOUND IS IN ESSENCE A RAISED DRAINFIELD. 

The mound consists of a sand bed over lightly tilled native soil, a gravel distribution bed on top of the sand bed, and a cover of topsoil over the entire mound. Septic tank effluent is distributed evenly over the sand bed through a pressurized network of small-diameter perforated pipes installed in the gravel bed. (Many of the early mounds that were not pressure-dosed failed.) Controlled dosing in the gravel bed aids in uniform distribution of the effluent into the sand. Final treatment and disposal occur in the native topsoil beneath the built-up mound.

Problems with Mounds

- They are huge (about 40 feet wide by 80 feet long for a 3-bedroom home with 450-gallon-per-day flow) and are an intrusion in the landscape.
- Routine monitoring and maintenance of the pump system is needed to ensure continued proper operation. When the power goes out, they stop working.

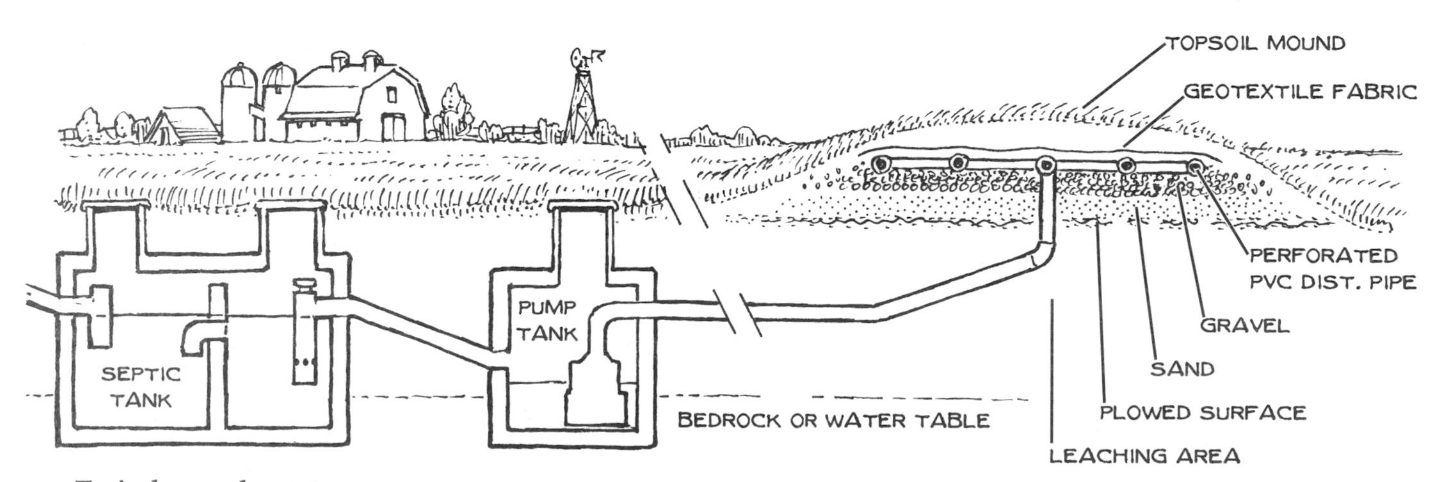

Typical mound system
Effluent flows from septic tank to pump tank, where it is pumped up to mound, which is in essence a drainfield raised above normal ground level.

- They are expensive (pump, pump chamber, pressure distribution network, clean graded sand, engineering fees, heavy equipment used in construction).
- When the soil is saturated from rainfall, they tend to leak around the toe.
- They are ecologically destructive. Large amounts of sand and gravel are mined from riverbeds and transported to the site. The land upon which they are constructed cannot be used for anything else.
- Careful design and construction is required. Mound system failures have occurred due to compaction of topsoil, improper configuration or orientation on sloping sites, undersized sand bed, a poorly designed or constructed pipe network, and sand material that is too fine or too coarse.

For further information on mound systems, review articles in *Small Flows Quarterly* and *Pipeline*, available on line and by mail from National Small Flows Clearinghouse at:

http://www.nesc.wvu.edu/nsfc/nsfc_index.htm

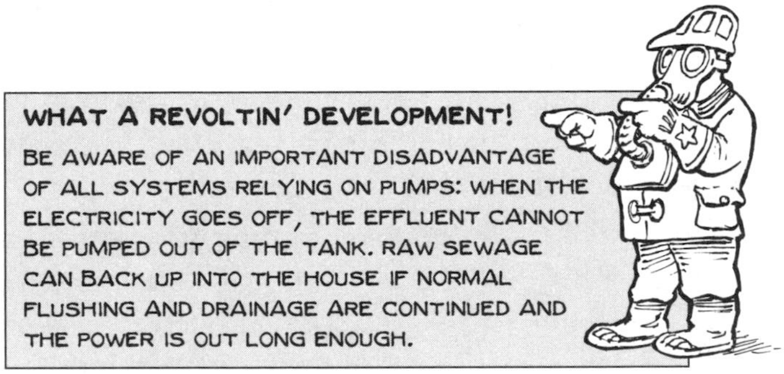

WHAT A REVOLTIN' DEVELOPMENT!

BE AWARE OF AN IMPORTANT DISADVANTAGE OF ALL SYSTEMS RELYING ON PUMPS: WHEN THE ELECTRICITY GOES OFF, THE EFFLUENT CANNOT BE PUMPED OUT OF THE TANK. RAW SEWAGE CAN BACK UP INTO THE HOUSE IF NORMAL FLUSHING AND DRAINAGE ARE CONTINUED AND THE POWER IS OUT LONG ENOUGH.

SAND FILTERS

There are three types of sand filters: the intermittent sand filter, the bottomless sand filter, and the recirculating sand filter.

Intermittent Sand Filter

This consists of a watertight basin with a bed of sand 24–36 inches deep with a distribution pipe network on top, and an underdrain below. The distribution network and underdrain system are laid out in the layers of gravel above and below the sand bed. A medium-grade, clean sand is used for the sand bed, and pea gravel is used for the distribution pipe and underdrain layers. *(See p. 92.)* The intermittent sand filter produces a high-quality effluent for disposal into difficult soil. It essentially serves as a substitute for treatment that would otherwise be achieved by good-quality drainfield soil.

The unit is usually constructed below ground, with a geotextile filter fabric over the top of the gravel distribution bed, and soil backfill up to native grade. However, some filters are built open to the ground surface with a removable wood cover or are constructed completely above ground in an enclosed structure similar to a raised garden bed.

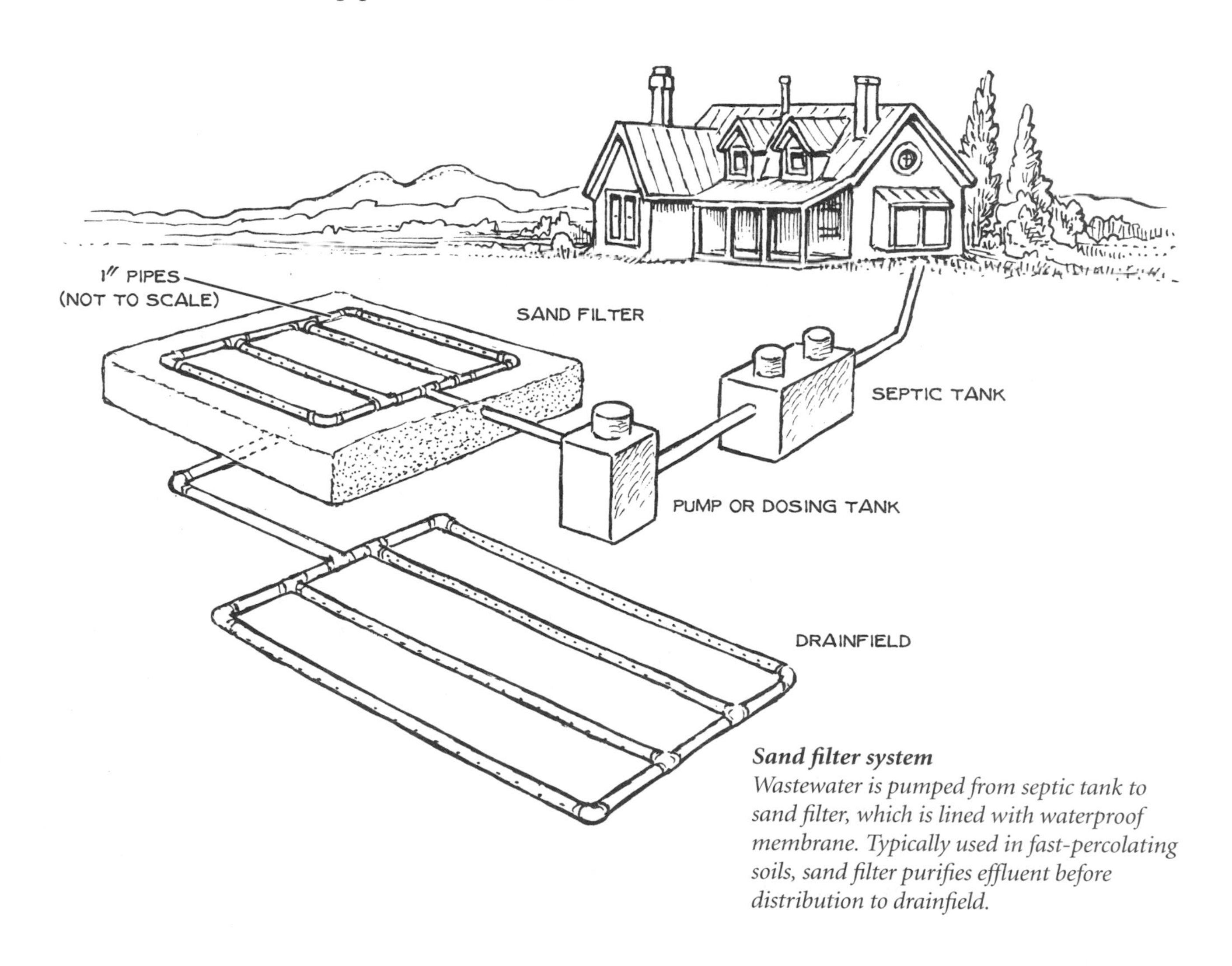

Sand filter system
Wastewater is pumped from septic tank to sand filter, which is lined with waterproof membrane. Typically used in fast-percolating soils, sand filter purifies effluent before distribution to drainfield.

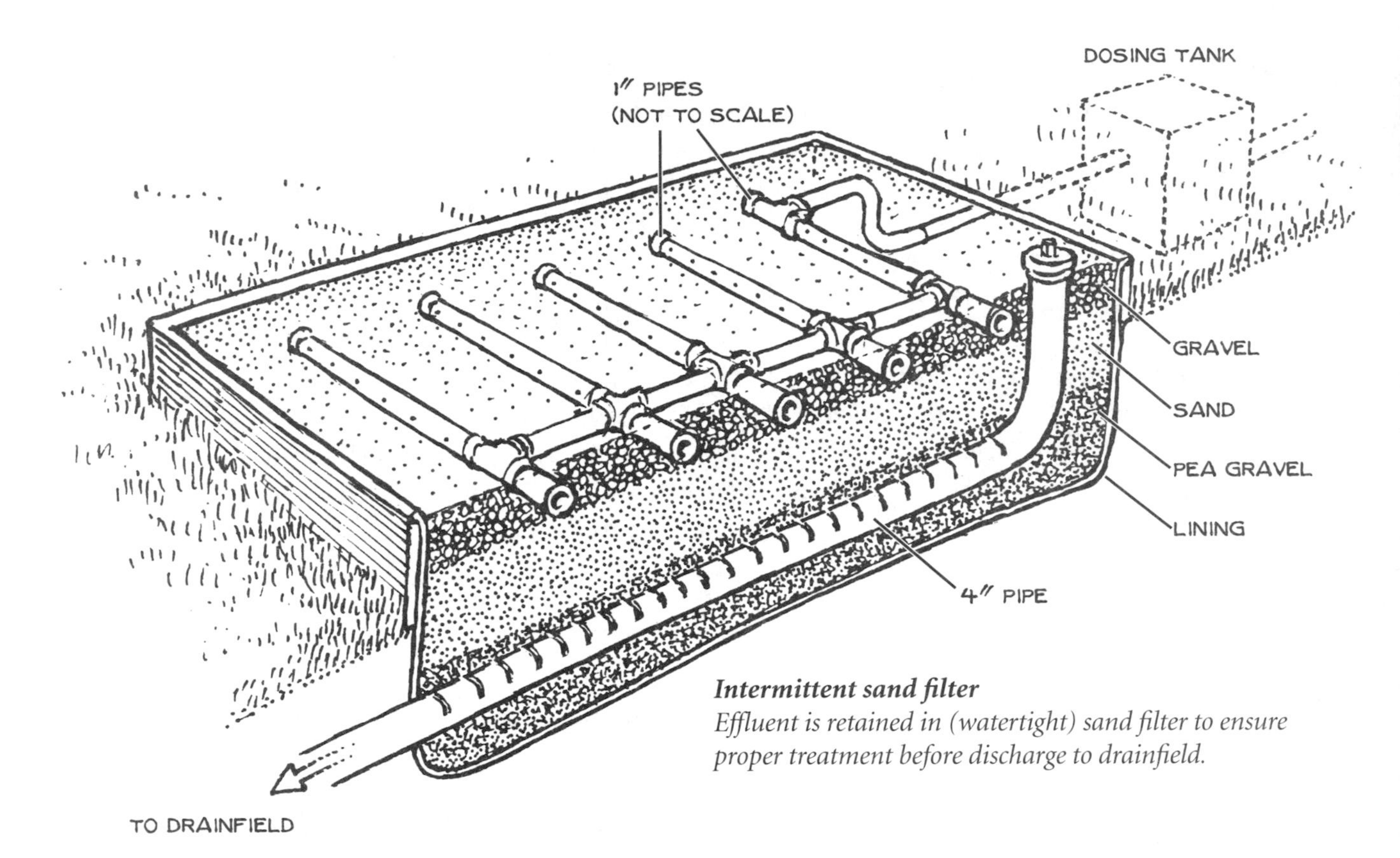

Intermittent sand filter
Effluent is retained in (watertight) sand filter to ensure proper treatment before discharge to drainfield.

How an Intermittent Sand Filter Works

Septic tank effluent is periodically (intermittently) applied to the sand filter through the pipes at the sand bed surface. The treated, filtered effluent is collected by the underdrain system below the sand bed, and then typically flows to the drainfield directly or via a second dosing tank for final disposal.

Size of Sand Filters and Drainfields

A typical 2- to 3-bedroom house (300-gallon-per-day flow) would use a sand filter unit about 19 feet by 19 feet and 4 feet deep. Since a sand filter produces better quality, clearer effluent than a septic tank, higher drainfield application rates may be allowed in some cases and a correspondingly smaller drainfield is possible. For further info, see:

http://www.epa.gov/OWM/mtb/isf.pdf

Also:
http://www.venhuizen-ww.com/

Bottomless Sand Filter

This is a sand filter that provides wastewater treatment not provided by excessively permeable (Type 1A) soil. This technology is being used for homes on sites where soil and site conditions preclude the use of conventional or shallow drainfields. There is a containment vessel consisting of four sides, but open at the bottom. Typical would be 12 inches of gravel with the distribution pipe, and under that 24 inches of sand or crushed glass; on top of the gravel is 6–12 inches of earth backfill. Pretreated effluent is applied via the pipes in the middle of the gravel layer. It then trickles down through the sand media in a time-dosed mode, and from there into the underlying soil.

See "NOPD II Helps Rhode Island Improve Coastal Pond," *Small Flows Quarterly*, Spring 2000 (Vol. 1, Issue 2), which describes bottomless sand filters and other methods of remediation in critical environmental areas; go to p. 10 of: www.nesc.wvu.edu/nsfc/pdf/SFQ/SFQsp00.pdf

Recirculating Sand Filter

A significant advance in recent years has been the development of recirculating filters with both sand and synthetic media. They are much smaller in footprint (for example, 4 × 8 feet) than a typical sand filter, and suitable for use in environmentally sensitive areas (sites on the coast, rivers, etc.). These devices recirculate effluent and provide a level of treatment superior to many centralized treatment plants at a fraction of the cost, especially when there is a necessity to remove nutrients like nitrogen.

The recirculating sand filter typically consists of three parts: a septic tank, a recirculating unit, and a drainfield, which can be a bottomless sand filter, shallow drainfield, or a standard drainfield). Effluent is recirculated between the recirculating tank and the sand filter, or drainfield, where it is applied in even doses. *Note:* Less land is required for a recirculating sand filter—one-fifth of the land area of an intermittent sand filter *(see p. 92)*.

For info on recirculating sand filters, see:

www.nesc.wvu.edu/nsfc/pdf/eti/RSF_tech.pdf
http://www.toolbase.org/Technology-Inventory/Sitework/recirculating-sand-filters

A SAND FILTER IS A GRAVEL-FILLED HOLE IN THE GROUND, LINED WITH A WATERTIGHT MEMBRANE, WHICH PURIFIES THE SEPTIC TANK EFFLUENT BEFORE IT GOES TO THE DRAINFIELD.

Sand-Lined Trenches

Sand-lined trenches are often used in excessively permeable (Type 1A) soils. Unlike an intermittent sand filter, they are not contained in a watertight vessel. Typical would be 12 inches of gravel with the pipe, and under that 24 inches of sand or crushed glass; on top of the gravel is 6–12 inches of earth backfill. Proper function requires that effluent to the sand filter be distributed over the media in controlled, uniform doses. To achieve accurate dosing, these systems require a timer-controlled pump, pump chambers, electrical components and distribution network, with a minimum of four doses per day spread evenly over a 24-hour period. The effluent is absorbed into the native soil at the bottom of the sand-lined trenches, which accomplishes disposal and further treatment.

For a good paper titled "Sand-Lined Trench Systems," which includes drawings of trenches and bottomless sand filters, by the Washington State Department of Health, see:

http://www.doh.wa.gov/ehp/ts/WW/SandLinedTr99.pdf

WHAT'S NEW IN THIS EDITION OF THE BOOK?

Following are some of the components and systems not covered in the first edition of this book six years ago. For one thing, new technology at the septic tank end of the system has proven to dramatically reduce solids loading on the drainfield, helping maintain healthy conditions in the soil. Aerobic treatment units, which can reduce the BOD (biological oxygen demand) in the septic tank effluent by over 90 percent, greatly reduce the load on the soil of the drainfield. Following are just a few advanced systems.

Trickling Biofilters

Cleaning up effluent: These are excellent for communities on bays, lakes, or rivers who are facing septic system upgrade requirements. Trickling biofilters are used in place of sand filters (and are much smaller) and are especially effective in nitrogen removal. They work by dripping effluent over a medium, which constitutes a large surface area where bacteria digest and purify wastes.

Maintenance must be performed regularly with these units, and many health departments require that a homeowner have a maintenance contract with a qualified inspection service. We list four different systems below. There are many other makes of trickling biofilters.

Advantex® Treatment System

The Advantex® System features a small (7½′ × 3′ × 2½′) fiberglass basin (it is small enough to fit under a deck) filled with a highly absorbent textile material that receives effluent from the septic tank. Bacteria on the geotextile fabric digest solids and produce a clear, odorless effluent that exceeds Secondary Treatment Standards. It can be monitored via telemetry by a remote monitoring system that will alert the homeowner to problems. It is produced by Orenco of Oregon.

http://www.orenco.com/ots/ots_adv_index.asp

Aerocell™ Treatment System

This system consists of several modular units that are configured according to specific site requirements, typically 4 to 6 modules for a 3- to 4-bedroom house, respectively. Effluent is sprayed over the surface of open-cell foam media and collected in the bottom of the modules for return to the septic tank for denitrification. The units are lightweight and can be moved by one person. Modular systems such as this offer flexibility of installation. Maintenance consists of regular inspection and cleaning of the spray nozzle, control panel, and pumps.

http://www.zabelzone.com/

Puraflo® Peat Biofilter

The Puraflo® is a modular pre-engineered biofiltration system that uses natural peat fiber as a biofiltering media. Primarily used in a single-pass mode, the processes occurring are filtration, absorption, adsorption, ion exchange, and microbial assimilation. The system significantly reduces biological oxygen demand (BOD) and total suspended solids. Ammonia and pathogens are reduced. These systems can be used in critical resource areas and used with shallow pressurized drainfields.

http://www.bnm-us.com/

Waterloo® Biofilter

This system was developed in Ontario for treatment of domestic wastewater and has been used for wastewater treatment at small towns,

> **NOTE:** THE EFFLUENT COMING OUT OF TRICKLING BIOFILTERS IS CLEAN ENOUGH TO GO DIRECTLY TO A DRIP IRRIGATION SYSTEM *(SEE NEXT PAGE)* IN ANY TYPE OF SOIL; NO DRAINFIELD REQUIRED.

resorts, and food processing facilities. It has been used in cold climates and on sites where water is reused onsite. The medium used is a lightweight open-cell foam which has a high surface area and is resistant to clogging. Settled wastewater is sprayed on the surface of the biofilter. The Waterloo® unit can be placed above or below ground according to site conditions and can be used in single-pass or multi-pass modes.

http://www.waterloo-biofilter.com/

Effluent Filters

These systems include pump vaults that filter and physically trap solids, either from a septic tank or a dosing tank, before they get to the disposal field. Water inside the septic tank (or pump vault) must pass through a screen before entering the disposal field. Screens come in a variety of mesh sizes. This is an inexpensive way to significantly improve effluent quality and thereby decrease solids accumulation in drainfields. *(For drawings of two types of filters, both inside septic tanks, see p. 9.)* Maintenance is important, since solids accumulate on the filtering screen and will eventually block it. Alarms can be installed to alert the homeowner when water level in the septic tank is rising and that the filter therefore needs cleaning. The filters are placed either within the tank or in a separate tank (pump vault).

Biotube®

This unit is manufactured by Orenco Systems, Inc. It is designed to be used in both new systems and retrofit applications and comes with filters with various size openings, and in diameters of 4, 8, 12, and 15 inches. See:

http://www.orenco.com/eps/eps_index.asp.

Septic Protector™

See p. 175.

www.septicprotector.com

Zabel Effluent Filters

www.zabelzone.com

Bio-Kinetic™ Effluent Filters

These go in a pump vault adjacent to the septic tank. They equalize flow and are rated for flow rates of up to 2,000 gallons per day.

www.norweco.com

Drip Irrigation Systems

Recycling water for the garden: Properly designed, drip irrigation systems are the most environmentally friendly drainfields, minimizing site impact and reusing water. Drip dispersal is useful in a number of soil conditions as it discharges into the shallow horizons where there is good percolative capacity, root uptake of water and nitrogen, more microbes, and high soil content. Treated and filtered wastewater is distributed directly into the soil slowly and uniformly from a network of narrow pipes. Wastewater is pumped through the drip lines under pressure and drips slowly from evenly spaced "emitters." There is minimal site disturbance due to the flexible tubing that can be placed around trees and shrubs. Drip systems must be carefully designed to follow the contours of the site as well as to avoid problems with drainback and freezing in the wintertime.

It's very important that the effluent be clean to avoid clogging of the emitter orifices. Some drip systems are designed for use with screened septic tank effluent. Wastewater is collected in a sump and pumped through the drip tubing.

Anti-siphon valves are placed at high points in the drip irrigation network to prevent backflow of soil particles into the emitters. A recent improvement in drip systems is recirculating the drip lines back through the system, which helps avoid clogging. For obtaining the National Small Flows Clearinghouse paper (#SFPLNL16) "Spray and Drip Irrigation for Wastewater Reuse and Disposal," see:

http://www.nesc.wvu.edu/nsfc/nsfc_pipelineposter.htm#sadi

Following are a few drip irrigation systems:

Wasteflow

Geoflow, Inc., is a large manufacturer of drip irrigation systems. Their drip systems include large orifices to prevent clogging. They also utilize what they call Root Guard technology, which releases an herbicide called Treflan® into the soil immediately surrounding the drip emitter to prevent root growth from entering the emitter. You certainly wouldn't want such a system near vegetables. The idea of using an herbicide in your garden seems questionable.

Bioline

This drip irrigation system has pressure-compensating emitters and a self-flushing feature to minimize clogging.

http://www.netafim-usa-wastewater.com/

Drip-Tech

These designs utilize Bioline (see above) and automatic backwashing disk filters.

http://www.drip-tech.com/

Perc-Rite

http://www.wastewatersystems.com/

Shallow Leachfield Systems

It is estimated that there are now over 750,000 shallow drainfields in North America, with more being installed each year. They can be installed with a shovel, requiring no heavy equipment, and there is no gravel. Most important, effluent is distributed in the top 16 inches of soil, where over 98% of soil life exists; very real purification takes place in this environment, as opposed to 2–3 feet down in a typical gravel-filled drainfield. In addition to the very popular Infiltrator® shallow drainfield chambers *(see p. 23)*, here are other manufacturers of chamber systems or pipe intended for shallow drainfields:

Cultec Contactor and Recharger

A variety of leaching chamber sizes

www.cultec.com/

Enviro-Septic®

Corrugated 12″ diameter drainage pipe wrapped with polypropylene mat and fabric.

http://www.PresbyEnvironmental.com/

Envirochambers

Chamber systems and corrugated drainage pipe wrapped with filter fabric.

http://www.hancor.com/

Restoring Failed Drainfields

If your drainfield has failed, or is showing signs of failure, first read pages 58–61, where partially failed and completely failed drainfields are described. Also, a helpful paper on the subject is "Drainfield Rehabilitation," the Winter 2005, Vol. 16, No. 1 issue of *Pipeline* from National Environmental Services Center:

http://www.nesc.wvu.edu/nsfc/pdf/pipline/PL_winter05.pdf

Following are two systems for drainfield rehabilitation that we have discovered since the first edition of this book:

The Pirana/Sludgehammer™ System

This consists of an "aerobic bacteria generation system," a unit mounted on the bottom of the tank near the inlet, and a 40-watt air pump at the top of the tank that aerates septic effluent. A special blend of microbes is introduced into the system which helps digest wastes in the tank. The microbes also travel into the drainfield, where they are said to digest the biomat and provide for freer distribution of effluent in the soil. Tests of the system at the University of California at Davis and approvals for remediation use by the Massachusetts Department of Environmental Protection show the Pirana's ability to rejuvenate otherwise failed drainfields. See:

http://piranaabg.com.

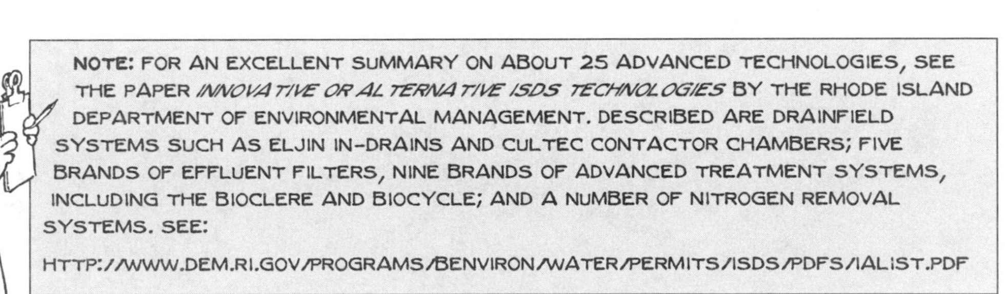

White Knight™ Microbial Innoculator/Generator

This unit consists of a high-density polyethylene cylinder, which is installed within the septic tank, continuously inoculating the tank with non-pathogenic bacterial cultures. An air pump provides fine bubble aeration and circulation within the system, bringing the bacteria into contact with fixed film substrate and suspended organic compounds in the tank. The bacteria digest organic wastes in the septic tank and in the drainfield.

www.knighttreatmentsystems.com.

WETLANDS

Wetlands provide a low-cost method of purifying effluent. They are typically used by municipalities, but there are a growing number of small scale wetlands being used for individual homes, providing final treatment of effluent where subsurface disposal methods cannot be used. The town of Arcata, California utilizes a freshwater wetlands marsh to treat the town's wastewater, and it is often studied and referred to as a working example of ecologically sensitive wastewater treatment.

Different Wetlands Designs

There are two basic types of wetlands design:

1. Free-water surface wetlands. These are shallow, open ponds with plants (such as cattails) rooted in the pond bottom and a free water surface open to the air.
2. Subsurface flow wetlands. These are shallow beds filled with gravel or sand in which plants are rooted. The top layer is earth, and water flows through the gravel or sand media several inches below the surface. You can walk on these.

Note: Transporting effluent directly from the septic tank to free-water surface ponds seems risky, since insects and animals have access and could transmit pathogens. A subsurface-type setup would seem safer. Free-water surface wetlands are generally not allowed for single family homes.

How Do They Work?

Effluent is piped from the septic tank into a man-made marsh rather than a drainfield. The

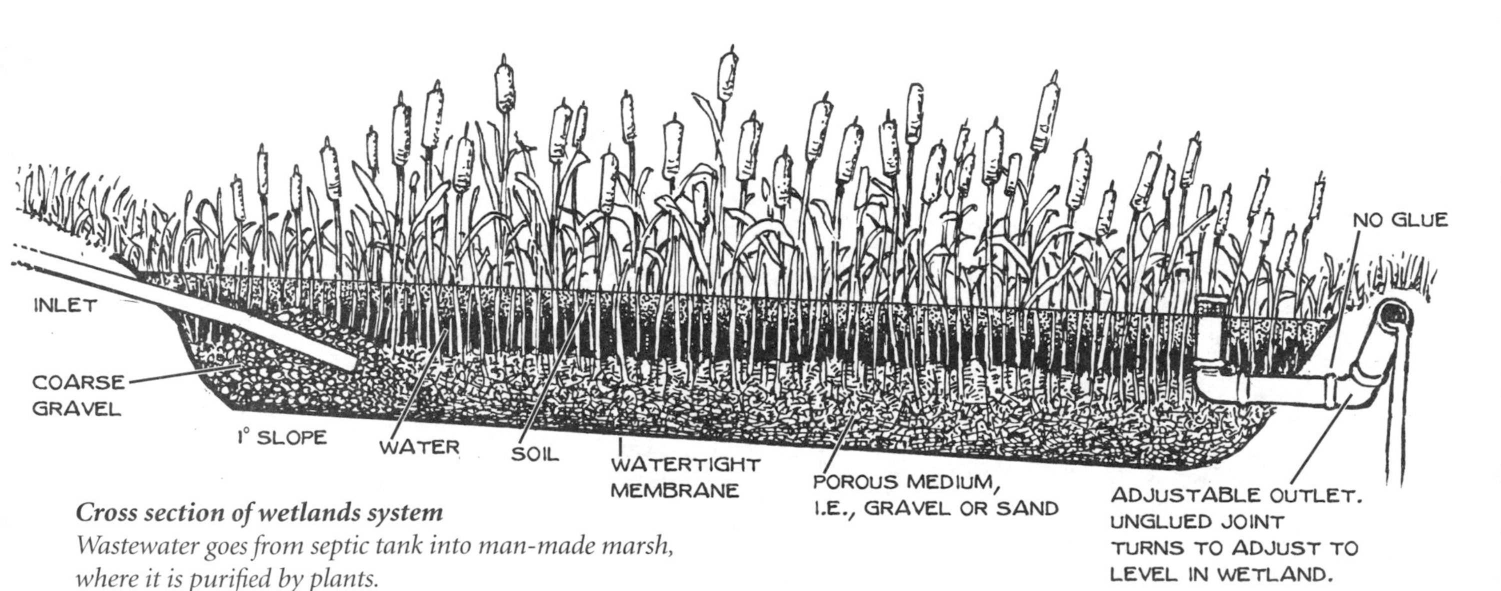

Cross section of wetlands system
Wastewater goes from septic tank into man-made marsh, where it is purified by plants.

CHAPTER TEN

Excessive Engineering and Regulatory Overkill

You might say it all started with the Clean Water Act of 1972, when billions of dollars were allocated to clean up America's water. With all that money floating around, it didn't take long for some engineers and some regulators to devise a methodology for extracting the maximum amount of grant money available. It was all so easy. First, septic systems are underground and out of sight; low visibility. Second, who could argue with the idea of "clean water"?

So 15–20 years ago, engineers and regulators (some of them) decreed that simple gravity fed septic systems were inadequate. They tightened up requirements, instituted new regulations, and thus began the new world of overblown, over-expensive septic systems. I got personally involved in a typical such scam in my hometown in 1989, and it was actually this experience (fighting against an albatross of a plan) that led to this book.

I considered writing about this situation when this book was first published in 2000. But the amounts of money were so huge, and the schemes so well orchestrated, I didn't think anyone would believe it. This was corruption completely missed by the media. The sums were huge. No one had an inkling. Well now, almost seven years later, the same things are going on, and more so. In this chapter, we'll give you the background, the history, and then case studies of small towns caught up in distorted engineering and excessive onsite wastewater disposal costs. In Chapter 11, "The Tale of Two Sewers," John Hulls describes how two California towns took two very different approaches in dealing with over-inflated wastewater projects. This leads into Chapter 12, "Small Town System Upgrades," where we describe how a community can take control of its own wastewater destiny and utilize local power in dealing with engineers and regulators.

A BRIEF HISTORY OF THE CLEAN WATER ACT AND THE EPA VALIDATION OF ONSITE WASTEWATER TREATMENT

There was very little national regulation of waste discharge until the Clean Water Act of 1972, which was spurred by the Cuyahoga River in Cleveland actually catching fire from pollutants, and the Santa Barbara oil spill, which coated the beaches of California with 250 million gallons of crude oil. The act required anyone discharging wastes to obtain a permit and properly treat the effluent from all point sources of pollution. It started a massive federal program to upgrade sewers and waste treatment plants, allocating over $250 billion by 1990. Citizens of Cleveland now enjoy dozens of dragon boats racing down the Cuyahoga River at the annual Burning River Festival, and many rivers and lakes in the nation have been restored to the point that the public can enjoy them once again.

marsh is sometimes lined with plastic or concrete. Water flows into another perforated pipe at the other end of the marsh and into an outlet tank, where a pipe controls the flow and keeps the effluent from surfacing in the marsh. Overflow from the tank goes through another pipe to a small pond where the water is "polished" by algae before evaporating or soaking into the ground (or being transported to a drainfield).

In cold climates, the reeds growing in the marsh die back in the winter. They form a thick blanket which insulates the marsh so that the roots continue to grow all year.

Purification by Plants

One interesting quality of wetlands is the ability of plants to assist in the cleanup of pollutants. Although the plants themselves take up very few of the pollutants, they provide oxygen to the bacteria—aerobic microbes that grow on root surfaces—that do most of the work. Anaerobic bacteria, which don't require oxygen, then finish the job. For example, dangerous nitrites are converted into nitrate, which is essentially plant fertilizer. It is also said that wetlands can clean up many hazardous chemicals that would not be filtered out in a regular septic system. For case examples of small community wetlands construction, see Natural Systems International at http://www.natsys-inc.com/. For an article on home-sized wetlands titled: "Build Your Own Constructed Wetland" from the *National Small Flows Journal*, see:

http://www.nesc.wvu.edu/nsfc/Articles/SFQ/SFQ_f04_PDF/Wetland_f04.pdf

A LARGE AND VERY COMPLETE STUDY COVERING OVER 100 WASTEWATER TECHNOLOGIES, PREPARED IN 2002 FOR THE CALIFORNIA STATE WATER RESOURCES CONTROL BOARD BY HAROLD LEVERENZ, GEORGE TCHOBANOGLOUS, AND JEANNIE L. DARBY, IS *REVIEW OF TECHNOLOGIES FOR THE ONSITE TREATMENT OF WASTEWATER IN CALIFORNIA*, AVAILABLE ONLINE (AS AN ACROBAT FILE) AT

HTTP://WWW.SWRCB.CA.GOV/AB885/TECHNOSITE.HTML

(THIS IS A VERY LARGE DOCUMENT TO DOWNLOAD, BUT AN IMPORTANT ONE FOR A COMPREHENSIVE OVERVIEW OF ONSITE WASTEWATER TREATMENT TECHNOLOGIES. YOU MAY WANT TO PRINT THE ENTIRE DOCUMENT AND THEN HAVE IT BOUND AT FEDEX-KINKO'S.)

10
Excessive Engineering and Regulatory Overkill

During this time, even though the emphasis was on providing treatment to major population centers, so much federal money was available that one EPA staffer said that if a kid in a remote part of Wyoming took a leak behind a bush, the EPA would receive a grant application to hook him up to a sewer. But it soon became apparent that centralized systems could not serve scattered rural populations economically, and even if the sewers and plants were funded by massive federal dollars, the operating expenses were way beyond what communities and individuals could pay.

Many people felt that the EPA, along with state and local officials, had ignored onsite treatment and forced communities into burdensome central systems. As housing development moved further away from urban areas, and there were problems with antiquated systems and direct discharge of waste to bodies of water, attention turned to onsite disposal. And as one seasoned observer said, "The EPA realized there wasn't enough money in the world to get everyone on to centralized sewers." Concurrently, the EPA started looking more closely at decentralized wastewater systems as a means of further improving the environment. (Up to that point the EPA had more or less considered septic systems an inferior form of wastewater disposal.) EPA researchers and regulators started serious studies of onsite treatment and new technologies as an environmentally favorable means of handling household waste.

In 1997, the EPA made a Report to Congress that recognized that onsite disposal systems can often protect the environment as well or better than centralized systems: "Response to Congress on Use of Decentralized Wastewater Treatment Systems" (EPA 832-R-97-001b. Date Published: 04/01/1997).

http://www.epa.gov/owm/mab/smcomm/scpub.htm

The executive summary made several key findings:

Benefits of Decentralized Systems

Decentralized systems are appropriate for many types of communities and conditions. Cost-effectiveness is a primary consideration for selecting these systems. A list of some of the benefits of using decentralized systems follows:

- *Protect Public Health and the Environment*—Properly managed decentralized wastewater systems can provide the treatment necessary to protect public health and meet water quality standards, just as well as centralized systems. Decentralized systems can be sited, designed, installed and operated to meet all federal and state effluent standards. Effective advanced treatment units are available for additional nutrient removal and disinfection requirements. Also, these systems can help to promote better watershed management by avoiding the potentially large transfers of water from one watershed to another that can occur with centralized treatment.
- *Appropriate for Low Density Communities*—In small communities with low population densities, the most cost-effective option is often a decentralized system.
- *Appropriate for Varying Site Conditions*—Decentralized systems are suitable for a variety of site conditions, including shallow water tables or bedrock, low-permeability soils, and small lot sizes.

- *Additional Benefits*—Decentralized systems are suitable for ecologically sensitive areas (where advanced treatment such as nutrient removal or disinfection is necessary). Since centralized systems require collection of wastewater for an entire community at substantial cost, decentralized systems, when properly installed, can achieve significant cost savings while recharging local aquifers and providing other water reuse opportunities close to points of wastewater generation.

With this change in perception at the federal level, along with an EPA-funded program for evaluating new technologies, many advanced systems were developed for onsite wastewater treatment and improved subsurface water infiltration systems (SWIS), the new name for the old drainfields. Since the first edition of *The Septic System Owner's Manual*, many of these systems are now in widespread use and approved by many state and local officials, not only for new installations, but the repair and upgrade of older septic systems. *(See Chapter 9, "Advanced Systems," pp. 86–99.)*

At the same time, health regulators have concluded that municipal sewage systems have significant problems, operating costs, and disadvantages, not limited to leaking pipes that directly contaminate the groundwater with raw sewage, infiltration problems, mixing with industrial wastes, concentrated environmental impact from high volumes of discharge at a single point, and dumping untreated waste during storm water overloads. All these factors are important considerations for a small community when weighing its options for onsite or offsite treatment of its wastewater.

"Stop Bad-Mouthing Conventional Onsite Systems"

This was written by Randy May, supervising sanitary engineer with Connecticut's Department of Environmental Protection:

"... Stop bad-mouthing conventional onsite systems. Our literature is replete with studies that illustrate that properly sited, designed, and installed conventional systems at rational densities are the most elegant and effective sewage disposal systems out there. Read the literature with care and understand that, when done properly, conventional septic systems are passive, cost-effective, tertiary systems. Recently, one author compared use of such systems with the long discredited practice of primary treatment/point source discharges to surface waters. That is a false analogy. A septic tank is a primary treatment tank where physical operations predominate. As a result, great process stability is the rule. The biomat in the leaching system and surrounding unsaturated soils structure provides highly stable secondary and tertiary treatment of effluent, virtually unmatched in sanitary engineering. ..."

21ST CENTURY SEPTIC SCAMS: HOMEOWNERS BEWARE!

As outlined above, the EPA began its "clean water" program by funding centralized sewers. They considered decentralized onsite wastewater treatment (septic systems) as low-tech and temporary. However, as time passed, the EPA realized there was not nearly enough money to have rural areas "sewer up." Concurrently, they began to look anew at septic systems and over the years have changed their outlook. Funding has been available now, for some years, from both federal and state sources, for small town wastewater improvements.

marsh is sometimes lined with plastic or concrete. Water flows into another perforated pipe at the other end of the marsh and into an outlet tank, where a pipe controls the flow and keeps the effluent from surfacing in the marsh. Overflow from the tank goes through another pipe to a small pond where the water is "polished" by algae before evaporating or soaking into the ground (or being transported to a drainfield).

In cold climates, the reeds growing in the marsh die back in the winter. They form a thick blanket which insulates the marsh so that the roots continue to grow all year.

Purification by Plants

One interesting quality of wetlands is the ability of plants to assist in the cleanup of pollutants. Although the plants themselves take up very few of the pollutants, they provide oxygen to the bacteria—aerobic microbes that grow on root surfaces—that do most of the work. Anaerobic bacteria, which don't require oxygen, then finish the job. For example, dangerous nitrites are converted into nitrate, which is essentially plant fertilizer. It is also said that wetlands can clean up many hazardous chemicals that would not be filtered out in a regular septic system. For case examples of small community wetlands construction, see Natural Systems International at http://www.natsys-inc.com/. For an article on home-sized wetlands titled: "Build Your Own Constructed Wetland" from the *National Small Flows Journal,* see:

http://www.nesc.wvu.edu/nsfc/Articles/SFQ/SFQ_f04_PDF/Wetland_f04.pdf

A LARGE AND VERY COMPLETE STUDY COVERING OVER 100 WASTEWATER TECHNOLOGIES, PREPARED IN 2002 FOR THE CALIFORNIA STATE WATER RESOURCES CONTROL BOARD BY HAROLD LEVERENZ, GEORGE TCHOBANOGLOUS, AND JEANNIE L. DARBY, IS *REVIEW OF TECHNOLOGIES FOR THE ONSITE TREATMENT OF WASTEWATER IN CALIFORNIA*, AVAILABLE ONLINE (AS AN ACROBAT FILE) AT

HTTP://WWW.SWRCB.CA.GOV/AB885/TECHNOSITE.HTML

(THIS IS A VERY LARGE DOCUMENT TO DOWNLOAD, BUT AN IMPORTANT ONE FOR A COMPREHENSIVE OVERVIEW OF ONSITE WASTEWATER TREATMENT TECHNOLOGIES. YOU MAY WANT TO PRINT THE ENTIRE DOCUMENT AND THEN HAVE IT BOUND AT FEDEX-KINKO'S.)

10 Excessive Engineering and Regulatory Overkill

CHAPTER TEN

Excessive Engineering and Regulatory Overkill

You might say it all started with the Clean Water Act of 1972, when billions of dollars were allocated to clean up America's water. With all that money floating around, it didn't take long for some engineers and some regulators to devise a methodology for extracting the maximum amount of grant money available. It was all so easy. First, septic systems are underground and out of sight; low visibility. Second, who could argue with the idea of "clean water"?

So 15–20 years ago, engineers and regulators (some of them) decreed that simple gravity-fed septic systems were inadequate. They tightened up requirements, instituted new regulations, and thus began the new world of overblown, over-expensive septic systems. I got personally involved in a typical such scam in my hometown in 1989, and it was actually this experience (fighting against an albatross of a plan) that led to this book.

I considered writing about this situation when this book was first published in 2000. But the amounts of money were so huge, and the schemes so well orchestrated, I didn't think anyone would believe it. This was corruption completely missed by the media. The sums were huge. No one had an inkling. Well now, almost seven years later, the same things are going on, and more so. In this chapter, we'll give you the background, the history, and then case studies of small towns caught up in distorted engineering and excessive onsite wastewater disposal costs. In Chapter 11, "The Tale of Two Sewers," John Hulls describes how two California towns took two very different approaches in dealing with over-inflated wastewater projects. This leads into Chapter 12, "Small Town System Upgrades," where we describe how a community can take control of its own wastewater destiny and utilize local power in dealing with engineers and regulators.

A BRIEF HISTORY OF THE CLEAN WATER ACT AND THE EPA VALIDATION OF ONSITE WASTEWATER TREATMENT

There was very little national regulation of waste discharge until the Clean Water Act of 1972, which was spurred by the Cuyahoga River in Cleveland actually catching fire from pollutants, and the Santa Barbara oil spill, which coated the beaches of California with 250 million gallons of crude oil. The act required anyone discharging wastes to obtain a permit and properly treat the effluent from all point sources of pollution. It started a massive federal program to upgrade sewers and waste treatment plants, allocating over $250 billion by 1990. Citizens of Cleveland now enjoy dozens of dragon boats racing down the Cuyahoga River at the annual Burning River Festival, and many rivers and lakes in the nation have been restored to the point that the public can enjoy them once again.

- *Additional Benefits*—Decentralized systems are suitable for ecologically sensitive areas (where advanced treatment such as nutrient removal or disinfection is necessary). Since centralized systems require collection of wastewater for an entire community at substantial cost, decentralized systems, when properly installed, can achieve significant cost savings while recharging local aquifers and providing other water reuse opportunities close to points of wastewater generation.

With this change in perception at the federal level, along with an EPA-funded program for evaluating new technologies, many advanced systems were developed for onsite wastewater treatment and improved subsurface water infiltration systems (SWIS), the new name for the old drainfields. Since the first edition of *The Septic System Owner's Manual*, many of these systems are now in widespread use and approved by many state and local officials, not only for new installations, but the repair and upgrade of older septic systems. *(See Chapter 9, "Advanced Systems," pp. 86–99.)*

At the same time, health regulators have concluded that municipal sewage systems have significant problems, operating costs, and disadvantages, not limited to leaking pipes that directly contaminate the groundwater with raw sewage, infiltration problems, mixing with industrial wastes, concentrated environmental impact from high volumes of discharge at a single point, and dumping untreated waste during storm water overloads. All these factors are important considerations for a small community when weighing its options for onsite or offsite treatment of its wastewater.

"Stop Bad-Mouthing Conventional Onsite Systems"

This was written by Randy May, supervising sanitary engineer with Connecticut's Department of Environmental Protection:

". . . Stop bad-mouthing conventional onsite systems. Our literature is replete with studies that illustrate that properly sited, designed, and installed conventional systems at rational densities are the most elegant and effective sewage disposal systems out there. Read the literature with care and understand that, when done properly, conventional septic systems are passive, cost-effective, tertiary systems. Recently, one author compared use of such systems with the long discredited practice of primary treatment/point source discharges to surface waters. That is a false analogy. A septic tank is a primary treatment tank where physical operations predominate. As a result, great process stability is the rule. The biomat in the leaching system and surrounding unsaturated soils structure provides highly stable secondary and tertiary treatment of effluent, virtually unmatched in sanitary engineering. . . ."

21ST CENTURY SEPTIC SCAMS: HOMEOWNERS BEWARE!

As outlined above, the EPA began its "clean water" program by funding centralized sewers. They considered decentralized onsite wastewater treatment (septic systems) as low-tech and temporary. However, as time passed, the EPA realized there was not nearly enough money to have rural areas "sewer up." Concurrently, they began to look anew at septic systems and over the years have changed their outlook. Funding has been available now, for some years, from both federal and state sources, for small town wastewater improvements.

During this time, even though the emphasis was on providing treatment to major population centers, so much federal money was available that one EPA staffer said that if a kid in a remote part of Wyoming took a leak behind a bush, the EPA would receive a grant application to hook him up to a sewer. But it soon became apparent that centralized systems could not serve scattered rural populations economically, and even if the sewers and plants were funded by massive federal dollars, the operating expenses were way beyond what communities and individuals could pay.

Many people felt that the EPA, along with state and local officials, had ignored onsite treatment and forced communities into burdensome central systems. As housing development moved further away from urban areas, and there were problems with antiquated systems and direct discharge of waste to bodies of water, attention turned to onsite disposal. And as one seasoned observer said, "The EPA realized there wasn't enough money in the world to get everyone on to centralized sewers." Concurrently, the EPA started looking more closely at decentralized wastewater systems as a means of further improving the environment. (Up to that point the EPA had more or less considered septic systems an inferior form of wastewater disposal.) EPA researchers and regulators started serious studies of onsite treatment and new technologies as an environmentally favorable means of handling household waste.

In 1997, the EPA made a Report to Congress that recognized that onsite disposal systems can often protect the environment as well or better than centralized systems: "Response to Congress on Use of Decentralized Wastewater Treatment Systems" (EPA 832-R-97-001b. Date Published: 04/01/1997).

http://www.epa.gov/owm/mab/smcomm/scpub.htm

The executive summary made several key findings:

Benefits of Decentralized Systems

Decentralized systems are appropriate for many types of communities and conditions. Cost-effectiveness is a primary consideration for selecting these systems. A list of some of the benefits of using decentralized systems follows:

- *Protect Public Health and the Environment*—Properly managed decentralized wastewater systems can provide the treatment necessary to protect public health and meet water quality standards, just as well as centralized systems. Decentralized systems can be sited, designed, installed and operated to meet all federal and state effluent standards. Effective advanced treatment units are available for additional nutrient removal and disinfection requirements. Also, these systems can help to promote better watershed management by avoiding the potentially large transfers of water from one watershed to another that can occur with centralized treatment.
- *Appropriate for Low Density Communities*—In small communities with low population densities, the most cost-effective option is often a decentralized system.
- *Appropriate for Varying Site Conditions*—Decentralized systems are suitable for a variety of site conditions, including shallow water tables or bedrock, low-permeability soils, and small lot sizes.

Unfortunately, the availability of clean water grants has created a "pork barrel" industry in various parts of the country, where unnecessarily expensive and ecologically disruptive wastewater plans are being forced on unsuspecting communities. Not only small towns, but individuals as well, now have to cope with overly restrictive regulations, and the cost of septic systems has skyrocketed.

It may have started 15 years ago in California, when engineers began to convince health officials that the tried and tested gravity-fed septic system did not work (at least in a majority of cases). Where previously you could install a gravity-flow system, homeowners now had to install high-tech, expensive, electricity-powered "mound" systems. My septic system cost less than $3000 in 1972, and it has worked beautifully for 34 years now. My neighbor, maybe 500 feet away, with the same soil profile, recently had to install a $40,000 mound system.

Homeowners Be Forewarned

There are two ways that you, the homeowner, will run across this scam:

1. *As an individual:* Either your system fails, or you build an addition, or some bureaucratic requirement of some sort means you have to hire an engineer who will design an expensive (where I live now $30,000+) system requiring a huge mound, pumps, and electricity in lieu of a simple, gravity-powered septic system which, in many if not most of these cases, would work fine.
2. *As a member of a community:* If this has happened, or is happening in your town, you'll see the following *modus operandi* in operation. If it hasn't happened to your town yet, watch for it on the horizon. Here's how it works:

Multi-Million Dollar Wastewater Plan

You and your neighbors will suddenly be confronted by a consortium consisting of engineers, regulators, and special interests who have devised an expensive community-wide plan. You will be told that:

1. *"Your town is polluting."* Testing shows a high coliform count (or high nitrogen levels) in the river, creek, lagoon, or groundwater. Invariably officials will not have conducted the DNA testing necessary to determine if the pollution is of human or animal origin. The pollution is assumed to be human, and *assumed* to be the result of failing septic systems. This is bad science, and many systems have been condemned on this basis. (Formerly officials said DNA testing was "too expensive," but it has come down in price in recent years.)
2. *"Your systems are failing."* Standards (often set by local engineers) will deem that your system is failing. "No, I'm afraid a gravity system can't work here and, by the way, you can hire me for $8000 to design a high-tech system."
3. *"We can get you a grant."* Grants are available for "clean water," and you will be told by regulators and engineers that your area qualifies for funding.

Typically this will take you by surprise. No well-publicized town meetings, just a small group quietly doing the planning. By the time you find out about it, ". . . the train has left the station." A building moratorium is sometimes enacted, to force residents to go along with an expensive wastewater system. Homeowners may be told they will not be able to sell their house if the area is condemned: scare tactics that are especially effective with old folks.

MY PERSONAL EXPERIENCE WITH A PORK-BARREL COMMUNITY WASTEWATER PLAN

In 1989, my town (Bolinas, California) got right to the brink of a publicly funded $7,000,000 septic plan for 300 houses. This plan was orchestrated by the county health department (as it was constituted in those days), an engineering firm hired by the county, and a small group of local people.

It was a poorly conceived and insensitive plan that was eventually abandoned, but only after an enormous amount of homeowner resistance. Grant money was available from "clean water" funding. In order to qualify for the grant, the county had declared Bolinas septic systems to be malfunctioning and by the time homeowners became aware of what was happening, the county, the engineers, the State Regional Quality Control Board, and eight or so Bolinas people were well on their way in design of a plan that would upgrade all 300 houses. (There were maybe ten actually failing systems.) A group of us started having Saturday morning meetings. From the outset we saw that the project was flawed in a number of ways. It was not based on science, practicality, or efficiency, but rather on obtaining grant money. As we studied the plan, we came to the following conclusions:

1. There was no proof that the purported pollution was human in origin. We wanted testing done to confirm the source: animal or human. It was never done.
2. We had determined, from our studies, that simple gravity-fed systems were, for the most part, working well in town.
3. We wanted nothing to do with any of the engineering firm's plans, which involved major disruption of the landscape, unnecessarily expensive high-tech "upgrades" for all 300 houses, and seizing land by eminent domain, if necessary.
4. Maintenance of these systems was going to be ongoing and expensive.
5. The county and engineers were desperate to get the grant money. When disposal fields could not be found outside the town, the engineers drew up a plan for "community leachfields," to be interspersed around the one-square-mile town. One resident was going to have effluent from 20 houses pumped to the lot next door.

To show people in town where this was leading, our group performed some guerilla theater: we took yellow construction tape out and on a Sunday morning roped off the proposed leachfield locations, putting up signs saying "Community Leachfield." People were astounded once they saw what was being planned.

Meeting after meeting took place for more than a year. The town's small newspaper was filled with a few pros and mostly cons. Cartoons were drawn. People were passionate. This wasn't right!

In the process, I got an insight as to how many grants work. Systems that work fine are deemed to be polluting in order to qualify for grant money. It was the first time I heard the phrase "non–cost-effective," which means that the engineers operate on the premise of: "How can we pick the most expensive solutions to use up all available grant money?" There was also the ethical question of grant money (public funds) being funneled to private homeowners: why should taxpayer money be going to upgrade personal homes of people already in an upper income level?

We eventually had a protest petition signed by 450 townspeople. In spite of this, the county and the engineers continued to push for the plan, but the funding had run out. Nothing was ever done. Predicted disasters never occurred. Most of the functioning gravity-fed septic systems continued functioning. There have been no health problems. Since then, however, any new construction or remodels require very expensive ($30-50,000) mound systems. (The county health department is currently far more open to homeowner [and scientific] input than it was 15 years ago, but it is still saddled with unreasonably restrictive standards.)

The engineers ended up collecting some $500,000 for designs that the town rejected. $500,000 would have easily fixed every truly failing system in the town and installed road drainage—which is actually our biggest problem—with enough left over to fund future repairs. It was a tragic waste of taxpayer money.

By the way, this book was generated by our experience in fighting this plan. We found there was no good book on septic systems for laypersons, so we published one.

Our experience showed us how it is possible for engineers and wastewater regulators, along with special interests, to misuse funding from the clean water legislation, ignore the wishes of homeowners, and craft bloated, expensive, and unnecessarily high-tech systems that generate maximum income for the engineers and health departments. Sad, but true.

The crudeness of these pork-barrel schemes is astounding. Upon examination, it's obvious that things are wrong. There's probably so little homeowner resistance because people don't understand septic systems. They're underground, invisible, and work so well that most people are unaware of them. When regulators and engineers appear, presenting a problem, along with a publicly funded solution, people fall for it. The planners are, after all, "experts." In fact, the plans make no sense other than generating maximum income for engineers, regulators, and developers.

The Risk Factor

One thing to consider when choosing a system is that there is no such thing as a no-risk decision, be it between central sewers or onsite waste treatment. While some people claim that individual systems are risky, the Center for Disease Control (CDC) statistics show virtually no cases of disease directly attributable to failing septic tanks. It is true that failing and improperly placed septic systems can contaminate wells and drinking water systems, but the risk of your septic tank failing and directly causing sickness is diminishingly small and is far less than the risk of being exposed to pathogens from a failing municipal sewer that can leak hundreds of thousands of gallons of raw sewage into public waterways.

The EPA has fined hundreds of municipalities for failures of this type, and many cities are facing costly renewal of aged sewers that are leaking hundreds of thousands of gallons into the groundwater. Onsite waste treatment and systems that pump treated effluent from advanced onsite systems pose far less human and environmental risk than large central collection systems, where failure of a single pumping station or breaking of a sewer main by earthquake, landslide or erosion can spill vast quantities of waste in a few minutes, or slowly leak into the groundwater when not maintained.

A Few Other Case Studies

In recent years I have seen this scenario in several other towns in Northern California. I've also heard of it in other parts of the country *(see below)*, and I suspect it's a growing phenomenon. There's just too much money to be made.

Monte Rio, California

The town of Monte Rio, California, on the Russian River, is being forced into a plan based on faulty science, railroaded through by a county supervisor and business interests, and entailing an engineering plan that is ludicrous and irresponsible. With no evidence of human pollution, residents are faced with a $10 million plan that calls for discharging sewage in a beautiful meadow (which is in a flood plain) on a local ranch. The meadow, on the banks of the Russian River, is to be taken from ranch owners by eminent domain.

For a very complete account, see:
http://shelterpub.com/_sepgaz/Monte_Rio.html

Los Osos, California

The town of Los Osos, California is (as of this writing) a textbook study of everything that can go wrong with wastewater funding. From *The Rock,* a small newspaper in Los Osos:

"The (state and local water) boards used invalid, manufactured data to support establishment of a prohibition zone that many believe is illegal, used information taken out of context to support claims that septic systems are not effective in wastewater treatment, and participated in the funding of an illegal loan to finance an over-priced, unnecessary sewer that the majority of citizens did not want. When citizens fought back, the local board attacked with cease-and-desist orders, in an attempt to intimidate its victims into submission."

Here is an excerpt from an email from Ed Ochs, Editor of *The Rock*:

"After what seems like centuries fighting to custom-build a STEP-collection system that fit the unique character of Los Osos, located in the Morro Bay watershed, a back-room deal to return to building a $220-million mega-sewer in the middle of town, in the heart of the community, was orchestrated by the Regional Water Board to overturn an election that moved the sewer out of town. It's been nothing but scam after scandal, but since Los Osos operates in a virtual media vacuum on the Central Coast, in a no man's land halfway between Los Angeles and San Francisco, our S.O.S. can only be heard on the quietest of nights.

"Because the county only builds big, expensive gravity sewers, the projected sewer bill for homeowners will range between $300 to $500 a month—which will wipe out one-third of the town's 14,000 residents—those who can least afford it, of course. It's an outrageous price tag for a mega sewer we don't need or want. But you've seen this movie before."

For further information on Los Osos (and an example of a small-town newspaper shining journalistic light on corruption) see:

http://rockofthecoast.com/

Nor is it limited to California:

Emmaus, Pennsylvania

"Briefly, without documenting the need through stream and well water samples, the township is trying to get 300 residents to pick up the entire $7.2 million cost of installing a gravity sewer, mostly in a high water table area next to a high-quality trout stream. Our engineer says 'It's gonna leak like a sieve' in five years and really pollute the stream the township claims it's trying to protect."

"A gravity sewer would open the door for massive new development in our rural community in one of the hottest housing markets in the country. Residents are pushing for a low-pressure system, which would severely limit future growth. It's a mess!"

–George DeVault

Wisconsin

This from Tom Larson, of the Wisconsin Realtors Association: "In preparation for the full implementation of the new changes to the state's private septic code (COMM 83), many counties throughout the state are in the process of revising their septic system codes . . . Many of these codes contain onerous soil inspection requirements that will cost homeowners thousands of dollars. One county adopted a soil inspection requirement that required a backhoe to dig soil borings every time a house with a septic system was sold or remodeled, even if the house sold multiple times during the same year."

TIPS FOR HOMEOWNERS

Homeowners Checklist

- Organize!
- Reign in engineers. Are they civil engineers (trained in compacting soil for building foundations) or do they have a degree in soil science, which focuses on friable soil?
- Don't trust "experts." Do the engineers have a conflict of interest? Have they had a hand in preparing local codes? Are they advising county health inspectors? Will they be designing an overall plan, and then be hired to design individual systems within the plan? There are engineers who favor expensive systems because their fees are that much higher.
- Insist that DNA testing be done.
- Educate yourselves. Study the alternatives.
- Communicate! Send out letters. Write articles for your local papers or letters to the editor. Inform others.
- Read pp. 117–132, "Small Town Septic System Upgrades." It takes you step-by-step through the process of hiring and controlling engineers and coordinating with public agencies.

State-of-the-Art Small Town Technology

In Oregon and Massachusetts, there have been extensive tests on advanced systems. In LaPine, Oregon, advanced systems have been tested for several years in an area with very shallow groundwater, where nitrates were causing health problems and degrading the quality of the Deschutes River, internationally famous for its trout and steelhead fishing.

See: http://marx.deschutes.org/deq/LaPineIndex.htm

Massachusetts has set up a test facility, MASSTC, in Barnstable County to certify systems for nitrogen removal and remediation of failed drainfields. They have approved systems to strip nitrates and preserve many small ponds such as Thoreau's famed Walden Pond. They have created a website which provides an excellent review of advanced onsite technology, including recirculating sand filters, peat filters, trickling filters, as well as graywater treatment and composting toilets; see: The Barnstable County, Mass. Alternative Septic System Test Center at:

http://barnstablecountyhealth.org/AlternativeWebpage/index.htm.

EPA Is Actually a Good Source of Info / *Small Flows* an Excellent Journal

As stated, the EPA has improved its attitude regarding onsite wastewater disposal in recent years. In fact the EPA is now more reasonable in its outlook than some states (like California), which have instituted regressive regulations requiring high-tech, expensive solutions.

In addition to funding the evaluation of new technologies, the EPA and Small Flows Clearinghouse put out a wide range of handbooks on how a small community can plan and manage septic systems in their area while avoiding the many costs and environmental disadvantages of centralized waste treatment.

As the EPA says in their onsite wastewater report to Congress cited at the beginning of this chapter, it's the way to go . . . (no pun intended).

The EPA Onsite Wastewater Treatment Systems Manual is at:

http://www.epa.gov/ORD/NRMRL/Pubs/625R00008/625R00008.htm.

Go here and click on "Chapter 4."

DNA Testing / Bacterial Source Testing

As stated before, it is highly important that a distinction be made between animal and human pollution. This type of testing is well described in the EPA "Wastewater Technology Fact Sheet" at:

www.epa.gov/owm/mtb/bacsortk.pdf

For example, several years ago this type of testing uncovered the fact that pollution formerly attributable to septic systems was in fact from raccoons:

DNA fingerprinting proved helpful when an oyster farmer on Virginia's Eastern Shore was faced with the closure of his shellfish beds due to elevated levels of *E. coli*. Failing septic tanks were assumed to be the primary source of the fecal pollution, but a survey of septic systems in the sparsely populated watershed indicated that they were not the cause, and it became necessary to identify other potential sources. The highest levels of coliform bacteria were measured in the small tidal inlets and rivulets of the wetlands located upstream of local houses, shifting suspected sources from human to other sources. Researchers collected fecal samples from raccoon, waterfowl, otter, muskrat, deer, and humans in the area and used DNA fingerprinting to confirm the suspicion that the source was not anthropogenic in nature. Comparing *E. coli* from the shellfish beds against the fingerprints of known strains in the DNA library, the researchers linked the in-stream *E. coli* to deer and raccoon (mostly raccoon). Several hundred animals, including 180 raccoon, were removed from areas adjacent to the wetlands. *E. coli* levels subsequently declined by one to two orders of magnitude throughout the watershed, allowing threatened areas of the tidal creeks to be reopened to shellfishing.

For a more complete story of the masked bandit incident, see:

www.bayjournal.com/article.cfm?article=547

BAD PLANS DESTROY COMMUNITIES

ONE OF THE MORE INSIDIOUS ASPECTS OF SOME COMMUNITY-WIDE PLANS IS THAT THEY FORCE RESIDENTS TO SELL AND MOVE OUT, RATHER THAN FACE THE INITIAL COSTS AND OPERATING COSTS ON A FIXED INCOME. IN FAR TOO MANY CASES, THESE "PLANS" OCCUR IN FORMERLY RURAL AREAS NOW WITHIN COMMUTE DISTANCE OF URBAN CENTERS. THE FORCED SALES BRING HOUSES ON THE MARKET TO THE BENEFIT OF THE REAL ESTATE INTERESTS. MANY A FORMERLY RURAL TOWN HAS BEEN "CARMEL-IZED," AS WITH THE MONTEREY PENINSULA IN CALIFORNIA, WITH AN INFLUX OF SECOND HOME OWNERS. THIS HAS HAD THE EFFECT OF DISPLACING OLD RESIDENTS AND RAISING RENTS FOR THE WORK FORCE IN THE COMMUNITY, OFTEN FORCING THEM TO COMMUTE LONG DISTANCES.

11
A Tale of Two Sewers

CHAPTER ELEVEN

A Tale of Two Sewers

by John Hulls

A precautionary tale for small communities in which we look at two very different ways to approach onsite waste disposal.

For the first group of folks it happened suddenly: regulatory agencies breezed into town, and announced that everyone was going to have to reduce the bacteria in the local bay by 75 percent . . . *or else!* And the agencies knew exactly who the culprits were, even before they did any tests: *septic tanks and farmers!* On one side of the bay, there was a small town and several houses strung right along the shoreline, some sitting on pilings over the water. Many of these houses had been there since the turn of the century so there were a lot of old septic tanks, and some houses had very limited land for the drainfields. But other homes in the community had adequate land for wastewater disposal, with septic systems apparently working fine. Regulators were talking about a major wastewater project, hooking up all the houses on the bay. Some people started asking questions . . . the local paper got interested. It would be the start of a long struggle.

In the other town, alongside a river, the first time most townsfolk heard about it was when they got notes in their mailboxes. After ten years of study by a small group of citizens, (including real estate interests), public officials (friends of the real estate interests), and engineers (who had already been paid hundreds of thousands of dollars), townspeople were finally called together and put things to a vote. Pollution had been detected in a stream leading into the river and even though no DNA testing was done to determine if the pollution was animal or human, they knew exactly what was causing the problem . . . *septic tanks*!

IS THERE REALLY A PROBLEM?

IT IS VERY IMPORTANT TO UNDERSTAND THAT THERE ARE POLLUTION PROBLEMS, AND BUREAUCRAT PROBLEMS. THE EPA WANTS ALL WATERWAYS TO MEET "FISHABLE/SWIMMMABLE" STANDARDS. IF A BODY OF WATER IS "IMPAIRED", THE STATES MUST PUT IT ON A LIST CALLED A 303(D), AND ESTABLISH A "TMDL" (FOR TOTAL MAXIMUM DAILY LOAD) FOR VARIOUS POLLUTANTS.

SOMETIMES, AGENCIES INVOLVED WILL SAY THERE IS A PROBLEM EVEN WITHOUT PROPER SCIENTIFIC PROOF BECAUSE IT ENABLES THEM TO GET REGULATORY AUTHORITY OVER AN AREA, OR EVEN WORSE, JUST TO GET FUNDING.

IF YOU'RE IN A SMALL COMMUNITY, DON'T TAKE THE WORD OF LOCAL REGULATORS AND ENGINEERS; YOU'LL OFTEN FIND THAT LOCAL AGENCIES ARE NOT FOLLOWING NATIONAL GUIDELINES. CHECK OUT NATIONAL STANDARDS AT THE EPA'S TMDL HOME PAGE:

HTTP://WWW.EPA.GOV/OWOW/TMDL/INTRO.HTML

Further, it was not a one-person/one-vote ballot. Businesses got more than one vote (such as ten for a restaurant), and the real estate developer who was part of the original group got 20 or 30 votes, one for each of his undeveloped lots. There were 200 undeveloped lots in all, each with a vote.

The plan proposed pumping sewage alongside a scenic river down a road prone to slippage and landslides, and spreading it on the meadow of a beautiful riverside ranch that had been in the hands of a ranching family for over a hundred years. Since the ranchers understandably didn't want to sell their land for sewage disposal, the county was prepared to take the ranch by legal action for eminent domain. By the way, the whole treatment plant would be in the flood plain of a river historically prone to overflowing its banks. After an initial meeting, many residents got upset. The local paper got interested. It would be the start of a long battle.

In the town beside the bay, homeowners were told that all their septic systems were failing. However, the county had a program where people could have their septic tanks inspected for free at no risk of penalty, so homeowners by and large did this. What the engineers had told them turned out *not* to be the case . . . only a handful had problems, many easily corrected. There were, however, a couple of very old ones right on the water that had to have their waste pumped and trucked away. After the survey, homeowners set about educating themselves about what was really going on with their septic systems. They formed a community study group designed to keep all property owners informed. They also took the important step of asking the county to pay for a project coordinator, who would work on behalf of the residents. They started to put together a plan.

In the town beside the river, things went from bad to worse. There had been absolutely no public outreach during the planning process, so the extent of the plans caught most of the townspeople by surprise. It looked like it was going to cost them tens of thousands of dollars to hook up to the new system, and around a hundred dollars a month per household for operating expenses. Homeowners were angry.

HOW DO YOU START A PLAN?

THE ENVIRONMENTAL PROTECTION AGENCY HAS PUBLISHED THE *VOLUNTARY NATIONAL GUIDELINES FOR MANAGEMENT OF ONSITE OR CLUSTERED (DECENTRALIZED) WASTEWATER TREATMENT SYSTEMS* AND THE *EPA SMALL FLOWS CLEARING HOUSE* PROVIDES SUPPORT FOR SMALL COMMUNITIES ON ALL ASPECTS OF WASTE TREATMENT, INCLUDING TRAINING PROGRAMS FOR BOTH CITIZENS AND ADMINISTRATORS. THEIR MAGAZINE, *SMALL FLOWS QUARTERLY*, SHOWS HOW HUNDREDS OF DIFFERENT COMMUNITIES HAVE DEVELOPED ENVIRONMENTALLY SOUND WAYS TO TREAT THEIR WASTE.

HTTP://WWW.EPA.GOV/OWM/MAB/SMCOMM/NSFC.HTM

At one meeting, flyers were distributed (it turned out paid for by the local real estate developer) scaring people about not being able to sell, remodel, or refinance their homes. There were handouts that slammed onsite treatment systems (directly contradicting Environmental Protection Agency documents). One flyer said that new sewer systems would be like a new car: "It smells good, feels good, and it's more reliable. The community will start to shine!" When a community member asked questions about the real need for the project and why the lack of scientific tests, the four presenters—the local supervisor, the engineer and two county bureaucrats—ducked every issue.

Back in the town by the bay, one of the first things the people did was to contact SFIC (Small Flows Information Clearinghouse). They started with a reprint from the SFIC magazine, *Pipeline*, the Winter 1997 issue, entitled "Choose the Right Consultant for Your Wastewater Project."

http://www.nesc.wvu.edu/nsfc/pl_1997.htm .

They paid special attention to the part in the introduction called "Remember—It's Your Project," which told them that it's up to the community to decide what it wants, and it can be disastrous if it's all left up to the consultants. After all, you wouldn't buy a car without figuring out what you really needed. The article also pointed out that it was essential to look at funding sources and work with them as the community decided what they wanted the system to do.

The community felt empowered by the article and the other things they learned. They decided they wanted the smallest feasible system, keeping as many individual systems as possible, and upgrading only those that were malfunctioning. Initially engineers had proposed a large collection and treatment system with every house in the community connected, but after looking at the alternatives, homeowners decided that what they really needed to do was to connect only those houses right next to the waters of the bay. They looked at many systems that had been installed in other counties and states. They also decided that they didn't want raw sewage pumped along the edge of the bay, so they specified that they wanted a system that pumped treated effluent from the septic tanks to a drainfield a safe distance from their bay. This is called a septic tank effluent pumping system (STEP system). One of the big advantages of a STEP system was that residents would not all be forced to hook up at once. It could be done incrementally. They set about working with the county to hire an engineer.

The town by the river didn't get to hire an engineer. The small group of people who started the project had already done so, and the one they had hired had run up close to a million dollars in engineering fees. Politically connected, the group got the county to impose a building moratorium so that not only would the project qualify for more federal and state money, but many homeowners would find it nearly impossible to upgrade, refinance, or sell their homes unless they went along with the plan.

The plan itself was horrendous—not just the costs. When the residents saw it, many were aghast. All the houses, businesses, and the vacant lots would be on a collection system that fed raw sewage into grinder pumps spread about town, and then pumped it down a narrow scenic roadway in a large pressure sewer to the beautiful meadow on the privately owned ranch previously mentioned. At the meeting to present the plan, some citizens asked what would happen during a power failure or a flood, and the engineer said that trucks with generators on them would go out to keep the pumps running. Some residents pointed out that the trucks would have had to drive though four feet of water in the last flood.

Some other people asked about how the project was getting funded. The county supervisor told everyone that the plan was in place and ". . . the train is leaving the station". Unfortunately, planning had taken so long that the railroad tracks no longer led to government funding dollars, but directly to the pockets of the town's homeowners and renters. Because of the way the county had locked itself into the funding, the option of a local community district wouldn't work. The only thing that the residents could do was try and change the rigged vote, and go after all the loose ends in the plan.

FINDING FUNDING

THERE ARE MANY RESOURCES FOR FUNDING FOR SEPTIC SYSTEMS. A GOOD PLACE TO START IS AT THE EPA'S SEPTIC SYSTEM HOME PAGE AT: HTTP://CFPUB.EPA.GOV/OWM/SEPTIC/HOME.CFM

CLICK THE LINK IN THE LEFT COLUMN FOR "GRANTS AND FUNDING." AS YOU CAN SEE, THIS SITE ALSO GIVES YOU ACCESS TO A WEALTH OF INFORMATION ABOUT ALL ASPECTS OF ONSITE WASTEWATER TREATMENT. START PLANNING FOR THE FUNDING AT THE VERY BEGINNING OF THE PROJECT.

BE SURE TO CHECK OUT THE NATIONAL SMALL FLOWS CLEARINGHOUSE SITE TOO.

HTTP://WWW.NESC.WVU.EDU/NSFC/NSFC_FAQ.HTM

Meanwhile, the bayside community was working with their plan. They went through every expense of the project. They decided they would allow homeowners to hire their own qualified installer to make repairs to their septic tanks and make the hookups. Not only did this allow the homeowners to get competitive prices, but since the engineers were only responsible for the *specifications* of the system, and not the installation, it saved tens of thousands of dollars in survey expenses, engineering drawings, and site supervision. The STEP system would consist of a pipeline running along the shoreline, collecting the treated effluent from septic tanks on lots that were too close to the water, and pumping it to a safe disposal site on a ranch. Out of some 70 houses along the bay, it turned out that they only needed to connect to 20, with the possibility of expanding the system to 38 later on. Many houses could use their existing septic tanks. Some of the other parts of the community would upgrade their septic tanks and drainfields, but most people's systems were just fine.

Better yet was the cost. By the time that they had worked out the most efficient contract, the entire engineering costs would be $500,000, and the total cost of the system would be $1.2 million. This compared to $25 million estimated before homeowners took over the project. Even better, they worked with their local supervisor and came up with grant funding for most of the project. As this is written, they'll be breaking ground on the project next year.

Back on the river, the angry citizens found all sorts of shenanigans with the funding, and the county was tying itself up in knots trying to figure out a way to hammer the project into something that might qualify for funding. The people who owned the ranch had repeatedly said that they were not going to allow their property to be used as a disposal site. The county, on recommendations from the engineers, assumed they could condemn the land, but when push came to shove, it wasn't going to cost the couple of hundred thousand dollars the engineers and planners had assumed, but millions.

The building ban that the regulators put in place is still stopping many homeowners from doing anything with their property. There is still no survey of the community septic tanks to see how many are actually malfunctioning. There has still been no DNA testing to determine if pollution is human or animal in origin. Incidentally, the river meets standards at the town, in spite of a tributary that is a major source of contamination several miles upstream. And now the community is facing millions of dollars of additional costs, and a plan is afoot to pump all the sewage upstream to a bigger town's treatment plant—a desperate measure. In fact, this plant has consistently failed to meet water quality standards during periods of high flow. The engineering fees are now over a million, and still rising.

ALWAYS DO A FULL SURVEY!

IT IS VITAL TO DO A GOOD SURVEY OF THE EXISTING SEPTIC TANKS BEFORE STARTING A COMMUNITY PROJECT. A PROGRAM LIKE THE MARIN COUNTY (CALIFORNIA) "SEPTIC MATTERS" PROGRAM THAT LETS HOMEOWNERS GET ANONYMOUS TESTS OF THEIR TANKS LETS THE COMMUNITY KNOW WHAT REALLY NEEDS TO BE FIXED, AND HOW BIG A PROBLEM EXISTS, OR IF THE SEPTIC TANKS ARE REALLY THE PROBLEM AFTER ALL.

WHY ARE "THEY" DOING THIS TO US?

MANY TIMES, THERE IS A LEGITIMATE REASON FOR UPGRADING WASTE TREATMENT SYSTEMS WHEN POLLUTION IS DOING DAMAGE TO A RECEIVING BODY OF WATER LIKE A RIVER OR BAY. HOWEVER, IN FAR TOO MANY CASES, NONEXISTENT HEALTH PROBLEMS ARE CITED, OR SEPTIC SYSTEMS BLAMED FOR BACTERIAL OR NUTRIENT CONTAMINATION WITH NO PROOF THAT THEY ARE REALLY THE CAUSE. ALL TOO OFTEN, A "PROBLEM" IS DECLARED SO THAT A PROJECT BECOMES ELIGIBLE FOR CERTAIN TYPES OF FUNDING.

WHEN THERE IS NOT ANY PUBLIC OUTREACH AT THE BEGINNING OF A PROJECT, ONLY SPECIAL INTERESTS, OFTEN ASSOCIATED WITH DEVELOPMENT, GET SERVED. FOR INSTANCE, LOTS CONNECTED TO A SEWER ARE OFTEN FAR MORE VALUABLE, ESPECIALLY WHEN THE REGULATORS OPPOSE SEPTIC TANKS. THERE ARE ALWAYS CONSULTANTS WILLING TO COLLECT FEES TO DESIGN AND REDESIGN, SEEMINGLY FOREVER.

12 Small Town Septic System Upgrades

CHAPTER TWELVE

Small Town Septic System Upgrades

Septic systems are failing. What can be done?

THE SCENARIO

You live in a small community, perhaps an unincorporated rural area. Many of the houses may be on septic systems built in the '40s and '50s and some of them are failing. The county health department may have responded to complaints and raised the red flag. Or the state may be pressing for a solution. There is a pollution control issue.

Town officials—the utility district directors, the mayor, other officials—are likely to be unpaid, part-time volunteers without technical skills in wastewater planning. You may be faced with a situation where you either take control of your own destiny, or state or county agencies will do it for you: hiring an engineer to design a solution, proceeding with construction, and assessing you for the result.

This chapter will give you some ideas about what will be necessary if you decide to deal with the solution within the community. Three major moves will be necessary:

- You will need a local agency authorized to deal with the project.
- You will need to educate yourselves in order to make informed decisions.
- You will need to hire an engineer.

LOCAL POWER

To sign a legally binding contract with an engineer, you'll need to set up a legal entity that can own, operate, and manage a wastewater facility. If your community is unincorporated, and you and your committee are private citizens, you'll have to decide when to establish such an entity. This is often a Catch-22 situation, because until the preliminary engineering reports are completed, you may not know whether a feasible plan or project can be developed.

Using an Existing Agency

It's easiest to utilize an existing agency such as a water district, granting them the authority to deal with the problem. Or, you may need to start by forming an informal property owners' association. Familiarize yourself with institutions already in place which deal with this problem. Consider the chain of enforcement from state to regional to local agencies.

Form a Special District if Necessary

If an agency isn't in place which has the authority to deal with the problem—to negotiate with other county, state, and federal agencies and to hire engineers and enter into contracts—you may want to form one. This may be a special district, a water district, or a wastewater district. Determine what your state requires to deal with sewage issues. This is a step that should be taken before extensive surveys or technical evaluations are done.

Not for the Faint-Hearted

Running a wastewater district is a serious undertaking: private citizens and neighbors (as elected officials) generally end up managing a sensitive and tricky situation. It's a lot different from running a water district, which tends to have few problems. Along with the obvious desirability of local control, there can be a downside. A newly created local agency may find itself faced with the difficult task of finding the right balance between regulatory overkill on the one hand, and irresponsible lack of enforcement on the other. Decisions formerly left to the more distant and less personal county or state agencies are now made in town.

If septic inspections will be required, how tough will you be in enforcement? Will you work with homeowners to allow the lowest-cost safe fixes, or will you take the easier way out of requiring high-tech (and expensive) repairs? Will elected volunteer officials take the time to deal with the individual idiosyncrasies of each site and work out a well-balanced solution? Will you be able to find the right personnel to do inspections and work with homeowners? What will it cost on an ongoing basis?

We suggest that you take the time to talk to other agencies in towns of your size. What has their experience been? What are their problems? What is their cost? If they had it to do over again. . . .

STUDY THE PROBLEM

This is a preliminary stage, to be followed later by a much more extensive study.

- Review existing working and non-working wastewater systems.
- Identify failing systems.
- Perform a door-to-door site survey if necessary.
- Study existing soil characteristics. *(See pp. 30–35 for some introductory information.)*
- Consider creeks, rivers, ocean, and natural drainage channels.
- Consider the potential for future growth and development.

STUDY POSSIBLE SOLUTIONS

Again, this is a preliminary study. The purpose is to help you understand the various options available, and the terminology. Studying both the problems and solutions at this early stage will prepare you for the most important decision you will make—choosing an engineer. In order to deal effectively with engineers you must know the language as well as the available options, so that you aren't dependent upon the engineer to decide what is appropriate for your community.

Using this book, learn what a septic system is and why things have gone wrong in your town. For example, was it uninformed household practices? Was it lack of maintenance? High groundwater? Poor soil conditions? All or a combination of the above?

Consider both the *collection alternatives*—gravity septic systems, pressure sewers, vacuum sewers, etc.—and *treatment systems*—standard drainfields, neighborhood treatment plant, constructed wetlands, etc. How will the solid waste be disposed of once it has been collected: recycled at a community facility which is part of the treatment process, or trucked to a landfill? (The latter option is becoming less viable all the time.)

Identify as many solutions as possible. A variety of technologies may be necessary for different problem areas. Study the single-household gravity-flow system first; then study the alternatives for individual systems, such as mounds and sand filters; and then the alternatives for group systems, such as STEP systems and sewers.

PAYING FOR IT

Funds tend to be limited these days, but loans might be available from the Farmer's Home Administration and possibly the EPA. Check with your county and state health agencies.

THE CONSULTING ENGINEER

The most important single decision you may make in solving your wastewater problem is the selection of a consulting engineer. We can't overemphasize the importance of taking enough time, care, and attention during the selection process to make sure you choose the engineering firm that's right for your community and its problems.

Before you begin the process, however, make sure that you and your steering committee, or planning committee, know your problems and also know what you want an engineer to do for you.

DEFINE THE PROBLEM, SEEK THE SOLUTION

People sometimes make the mistake of defining problems in terms of a particular solution. This can keep them from considering other solutions which may be better for their situation and less expensive. For example, if you have a water shortage and you say, "We need a new well," you are thinking of just one solution and may be overlooking other solutions, like fixing the pump, lowering the pipe, or practicing water conservation.

Start by facing the fact that you have a problem. This opens the way to thinking about alternative ways to solve it. Spend time as a group discussing the problem you want to solve.

Often, the best solution is a combination of several alternatives. Start by thinking about the various ways your problem can be solved.

Asking Questions to Help You Find Solutions

- What would be the best solution if money were no object?
- What would be the least expensive way to solve the problem?
- How many years will the solution last?
- Can we break the solution down into a series of affordable steps? Or would it be better to do it all at once?
- Is the solution so technical that we'll have trouble finding someone to operate and maintain it?
- Can we do all or part of the work ourselves?
- *What will happen if we don't do anything?*

Consider Costs of Various Solutions

- Have we considered the cost of operating the facility once it's built?
- How much would customers be willing to pay for service?
- Can we explain why improvements are important enough for people to go along with a rate increase?
- Will we need to borrow money, or can we self-finance?
- If we do need to borrow money, where can we borrow it from?

It's very important to have a sense of what you can afford—or are willing to pay for the facility—before you hire an engineer. Like asking a stranger to buy a car for you: if you can't give the engineer some idea of what you can afford, you may get a Cadillac (or Edsel!) solution when you can only afford a Volkswagen.

Once you have a clear sense of the problem you're trying to solve, some ideas about alternative solutions you want to consider, and a sense of what you'll be able to afford, you're ready to look for a consulting engineer.

WHAT YOU'LL NEED AN ENGINEER TO DO

Engineering services are generally divided into three parts:

Preliminary Engineering Report

In this first step, the consulting engineer examines alternative designs and costs, and recommends the best *affordable* approach to solving your problem. It should include a preliminary cost estimate for the recommended approach. The Preliminary Engineering Report is used as the basis for securing financing and further project development and must be approved by the regulatory agency.

Final Plans and Specifications

Once financing is arranged, you'll need to have your engineer prepare the Final Plans and Specifications. These are used to obtain design approval from regulatory agencies and are used by construction firms to bid on and guide construction of the improvements.

Construction Inspection

This is the final engineering phase: inspecting the work of the contractor you've hired. The purpose is to ensure that it's constructed according to the design specifications and also to approve any necessary changes. This might involve an engineer other than the designer of the plan.

Consider whether all phases of a project should be combined into one contract or if it is better to divide the work.

Before beginning the search for an engineer, you may want to create a selection committee of three people who will guide the selection process. *If you think you'll be seeking financing through a federal or state agency, contact that agency before you begin the selection process to ensure that you can satisfy that agency's requirements.*

PREPARING A REQUEST FOR PROPOSALS

Once you have determined the scope of the project, you will develop a request for proposals (RFP). Here you will be soliciting bids. Write out exactly what you want the engineer to do. Generally this will include:

- preparing a preliminary engineering report listing solutions to the problems you describe
- recommending the best possible solutions *(taking into account your stated financial limitations)*
- preparing a cost estimate
- helping you submit applications for financing if necessary
- doing final design and construction drawings once financing is secured
- (possibly) providing construction inspection services when the project is underway

Your Request for Proposals Should Contain These Key Points

- *Instructions*—Specify the due date, address for sending the proposal, length limitations, potential sources of funding, and format requirements.
- *Problem description*—Briefly describe your community, the problems you are facing, and your efforts to solve them.

- *Required services*—Indicate what you want the consulting firm to do.
- *Qualification*—Summarize the qualities you are looking for in an engineering firm, the designated project team, and particularly the team manager.
- *Reporting requirements*—State the kind of interim and progress reports you require.
- *Types of contracts*—Indicate the type of contract you desire; for example, fixed cost, cost plus fixed fee, or time and materials.
- *Criteria for evaluation and selection*—To ensure fair competition, describe how you will review and select the firm.

SELECTING THE ENGINEER

Make a list of five engineering firms that might be able to meet your community's needs.

Contact your state regulatory agency, county officials, or the Rural Development Administration and ask for some names of firms that have experience with small community wastewater problems. You might also ask nearby communities that have resolved similar wastewater problems for the names of their previous engineering firms. This is especially important if you're looking at some of the new, alternative technologies.

Many engineers have never designed an operational alternative system. Consult with state professional engineering associations and review advertisements in professional journals. Contact the National Small Flows Clearinghouse to obtain a list of firms with experience in the technology you are considering. *(See p. 127.)* Advertise your RFP (following your state's guidelines) in newspapers or journals. Check all references carefully and make sure that the systems designed by a prospective engineer actually perform as originally intended.

Contact all the firms on your list.

Let them know you're interested in contracting for engineering services and that you want a proposal outlining their qualifications. Again, if you plan to use state or federal money for the project, check with the funding agency to see if there are special requirements about securing engineering services.

Select the top three engineering firms submitting proposals.

Have committee members review each proposal *against the selection criteria* stated in your request for a proposal, and select the top three firms for interviews.

The proposals you receive in reply to your RFP should describe the following:

- the firm's understanding of the problem
- the firm's experience in small community wastewater planning and construction supervision
- the general approach to be used
- a description of each task, including estimated costs, timing, and staffing
- qualifications of the firm and project team
- a list of similar projects
- a list of references
- a resumé of each project team member
- a description of the firm's organization and resources relevant to the work

The proposal should indicate if hiring or subcontracting is necessary to undertake the work. The proposals should include a cost proposal that indicates specific cost (direct labor costs, overhead, travel costs, purchase of supplies and equipment), as well as determination of profit for the firm.

Check references of your committee's top choices.

Contact communities that are listed as references. Don't be afraid to ask tough questions. Below are some sample questions to ask when you call references:

- Were you satisfied with the quality of the work?
- Was the engineering firm able to meet the time frame and schedules agreed upon in your contract?
- Was the engineer willing and able to work closely with your community?
- Did the engineer keep the community periodically informed of the progress and direction of the project?
- Did the project stay within the budget, or were there unexpected additional costs?
- Did the engineer have other projects scheduled that caused time delays on your project?
- Did the engineering firm assist you in obtaining financing? Was this successful?
- Did you have any problems that would keep you from hiring the firm again?
- Is your system operating properly, and are you happy with its performance?

Eliminate any firms whose references don't check out.

Set up interviews.

Once you've pre-screened the proposals and checked the references, you'll want to have face-to-face interviews with each one of the firms. Allow enough time for each interview—about one hour. Set limits for the engineer's presentation so your committee has plenty of time to ask questions. (Require the engineer who'll be assigned to your project to participate in the interview.)

Use the interview to talk with each firm about your problem and strategies for solving it. Let them know you're looking for the best possible solution for the money allocated.

Let the engineers know what the financial limits of your community are, and make sure they understand that you're concerned about the long-term operating and maintenance costs, not just the initial capital costs. Tell them you don't want to have to raise rates continually to cover unexpected operating costs.

Tell the firm that your community doesn't necessarily want the most sophisticated solution. What you want is a system that's reliable, affordable, simple to operate, and economical to construct and operate.

Questions to ask

Prepare a list of questions ahead of time and ask each firm the same questions, such as:

- What experience does your firm have with projects like ours?
- Are you willing to look at reliable innovative and/or alternative designs in collection and/or treatment?
- Are there specific itemized services you don't provide?
- Are you familiar with various funding agencies? What has your experience been in working with funders? Has your firm assisted communities with grant writing and preparation of loan applications? What has been the success rate of those applications?
- Is the firm willing to enter into a fixed cost, "not-to-exceed" contract, or one in accord with the financing agency's fee schedule?
- Who, specifically, in your firm will be working directly with our board?

- What other projects are you currently working on that could take time away from our project?
- Will the engineer be willing to attend public meetings and discuss the project with customers?
- Will you agree to keep the community informed as planning progresses?

Make your final selection.

Have your selection committee evaluate all of the information that's been gathered. Discuss the pros and cons of each firm. Then use your best judgment and select the firm you believe will do the best job of solving your community's wastewater problem. Pick a firm that your committee trusts, and with whom you feel you can work.

Note: It's a good idea to maintain a written record that documents the basis for your selection.

CONTRACTING AND NEGOTIATING FEES AND PAYMENTS

As soon as possible, negotiate a contract and payment schedule. If you'll be financing your project through state or federal sources, then the financing agency will have guidelines. Many have standard engineering contracts and fee schedules you can use to complete this step. In other words, the form of contract and payment may be governed by the method you will use to finance the entire project.

Negotiating with an Engineering Firm

Negotiate carefully. Rely on written, rather than verbal, understandings. Be sure you understand what's in the contract before signing. Use an attorney to clarify vague sections. Negotiate services, timetables, and price. Check the reasonableness of the engineer's proposed fees with organizations like the Professional Engineers Council, the Board of Examiners for Professional Engineers, or your state Rural Development Administration office.

Finally, tie your proposed payments to tasks like completion and acceptance of preliminary engineering report, state approval of final design, completion of a given percent of construction, and final inspection. If you plan to borrow RDA funds to finance the project, ask the engineer to accept payment for the preliminary design report *when the project is funded.*

The governing body of the legal entity representing the community needs to review the final draft of the contract for engineering services and approve it before the officials of the governing body sign it and legally bind you to the terms of the contract. You're responsible for financing the project and paying for what gets done. Remember, you and your neighbors will have to live with your choices for a long time to come. Make sure you take the time to figure out how much you're willing to spend and what the best solution is, and follow the steps outlined to select an appropriate engineer for your project.

If you're not able to negotiate a satisfactory contract with your top selection, notify them in writing that you are breaking off negotiations —and begin negotiations with your second choice. Once you've reached a satisfactory agreement, notify the other firms that you have completed a contract with the selected firm.

An Engineering Contract — Minimum Requirements

The contract should set out a clear understanding of:

- services to be provided
- a timetable for completion of the engineering
- price for services
- payment schedule stating when you'll pay

Note: Price and payment schedules are especially important to small communities with limited resources.

Methods of Payment

Engineers sell their time and experience. They expect to be paid for the time and expertise they bring to your project. There are a number of approaches to paying engineers. They include:

- *Flat fee:* The engineer is paid an agreed-upon price for specified work.
- *Time and material:* Payment is based on the number of hours spent by the firm times an agreed-upon hourly rate schedule, plus out-of-pocket expenses. Often, payment is limited under this method through a *not-to-exceed clause.*
- *Percentage of construction costs:* The engineer is paid an amount equal to an agreed-upon percentage of construction costs.

THE IMPORTANCE OF ONGOING COMMUNICATION

A *very* important part of the process of designing a town septic system is that the engineer keep not only the committee members informed as to the progress of the project, but the general populace as well. Some engineers will write articles in the town paper to keep people up-to-date on what is going on.

An example of what can happen when there is lack of communication occurred in our small Northern California coastal town in the early '90s, when a septic project got all the way to the final stages without townspeople being aware of its size or technology. When the project was almost ready to start, and people became aware of the design and its implications, a petition signed by over 400 people convinced the utility district to abandon the project. This in spite of the fact that over $5 million in funding was available from the federal government. The result: over $500,000 of public funds spent on an engineering study that went nowhere. If the public had been informed in a clear manner early on, much time and money could have been saved.

Note: This book actually was born when a group of us, in battling the project, discovered that there was no good, non-technical book available for homeowners.

Bloated projects are not uncommon when public funding is available. Engineers have been known to over-design a project when substantial funds are available. In fact, there is a phrase, "non–cost-effective," used to describe solutions designed to use up available funding.

Ask your engineer to write periodic articles for townspeople, and be sure to have town meetings with the engineers present when critical decisions are to be made. Publish updates on a regular basis and ask for feedback.

DECENTRALIZED SYSTEMS IN SMALL TOWNS

The passage of the *Clean Water Act* in 1972 mandated that streams become swimmable and fishable and essentially mandated "sewerage" for everyone, a task which has proved fiscally impossible. Thus decentralized wastewater solutions are of critical importance to small towns being required to upgrade failing septic systems. Decentralized systems are more challenging than large "treat it, dump it" systems because more complex technology with more variables is involved. This trend is apparent in the new textbook for wastewater engineers, *Small and Decentralized Wastewater Management Systems*, by Ron Crites, M.S. and George Tchobanoglous, Ph.D. *(See bibliography, p. 166.)* This book covers a range: from disposal of one million gallons/day —about 10,000 homes—to individual onsite systems. Reclamation—not disposal—of water (including blackwater) has become one of the top priorities in thinking about the design of wastewater treatment systems. "Think of effluent as a resource. It's ridiculous to use drinking water to water the lawn," says Dr. Tchobanoglous, who goes on to note that changes are beginning to take place on a noticeable scale, citing a project in St. Petersburg, Florida with dual water systems —a system for drinking water and a system of reclaimed water for use in landscaping.

SMALL WASTEWATER SYSTEMS

Alternative Systems for Small Towns and Rural Areas

On the following four pages, 20 different wastewater options for small communities are shown. They vary from the simplest onsite systems, such as a septic tank and drainfield, to more complex offsite systems, such as the septic tank effluent pumping (STEP) system or small-diameter gravity sewers. This section is based on a flyer published by the Office of Water at the United States Environmental Protection Agency.

If your town is considering community-wide upgrades, looking through these pages will give you some ideas of the options available. The charts also provide easy-to-understand, thumbnail sketches of a variety of systems.

Why Small Systems?

Lower Water and Sewer Rates

If relatively few people have to pay for a large sewer system, rates skyrocket. Small, natural systems are simpler to operate and maintain, and make efficient use of the inherent capacity of the land.

Save Energy, Water, and Materials

Most small systems rely on gravity and natural biological processes in the soil to recharge groundwater. They use less mechanical equipment and energy, as well.

Save Farmland and Prevent Urban Sprawl

Large regional sewer systems in rural areas encourage growth and loss of open space.

Local System Management

The technologies shown here vary according to site, soil, population, and weather. Through local system management, small communities can choose the system most appropriate for their area, and maintain their rural character while developing a natural system that harmonizes with the environment.

Small Flows for Small Towns

Hats off to West Virginia University for the National Small Flows Clearinghouse, whose slogan is "Helping America's Small Communities Meet Their Wastewater Needs." They are a great resource for any small town facing septic upgrades, offering free and low-cost technical assistance, products, and information on small town onsite wastewater treatment and pollution prevention.

National Small Flows Clearinghouse
P.O. Box 6064
Morgantown, WV 26506
800-624-8301
http://www.estd.wvu.edu/nsfc

Technical Data

Engineers, consultants, and regulators can get more detailed technical information from the EPA's *Onsite Wastewater Treatment and Disposal Systems Design Manual, Septage Treatment and Disposal Handbook,* and *Alternative Sewer Systems Design Manual* from the Center for Environmental Research Information, 26 W. Martin Luther King Dr., Cincinnati, OH, 45268. 513-569-7562.

WASTEWATER OPTIONS FOR SMALL TOWNS

COMMON ONSITE SYSTEMS

1 Septic Tank & Gravel Absorption Trench

This is the most common system used on level land with adequate soil depth above the water table. Heavy solids in the liquid settle and greases float to the top of the tank. Bacteria break down some solids. The liquid flows from the tank through a closed pipe into perforated pipe and into gravel-filled trenches, where it seeps into the soil. Bacteria and oxygen purify the liquid as it slowly moves through the soil. Inspection ports permit checking liquid depth. Regular pumping of the tank reduces the solids discharged into the trenches and extends the life of the system. Using two-compartment septic tanks and resting the trenches (#4) are also recommended to extend trench life.

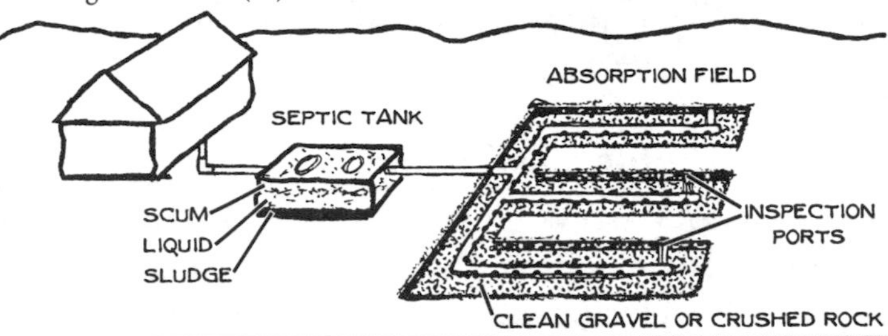

2 Septic Tank with Serial Distribution

Starting with the highest, each trench fills completely, then overflows through one drop box to the next. The effluent floods all soil surfaces. The drop box enables inspection of the system and control of discharge into each trench. Capping the pipe outlets in the upper trench forces resting. Serial distribution automatically loads upper trenches and minimizes the loading on lower trenches. Used on gently-to-steeply-sloped sites.

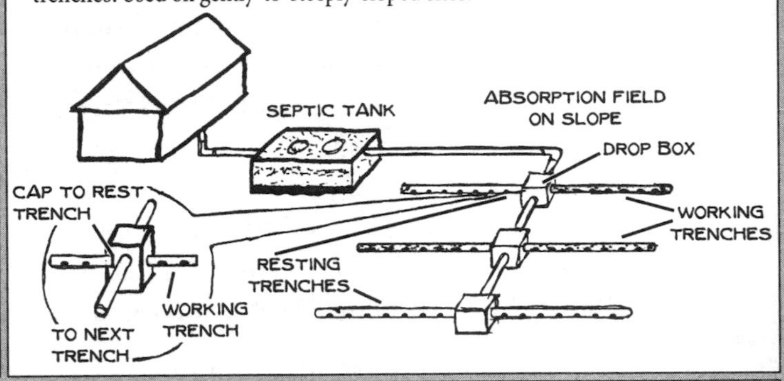

3 Septic Tank & Leaching Chambers

Open-bottom concrete chambers or arched plastic chambers create an underground cavern that stores effluent. The effluent floods the soil surface prior to seeping vertically through the bottom of the chamber.

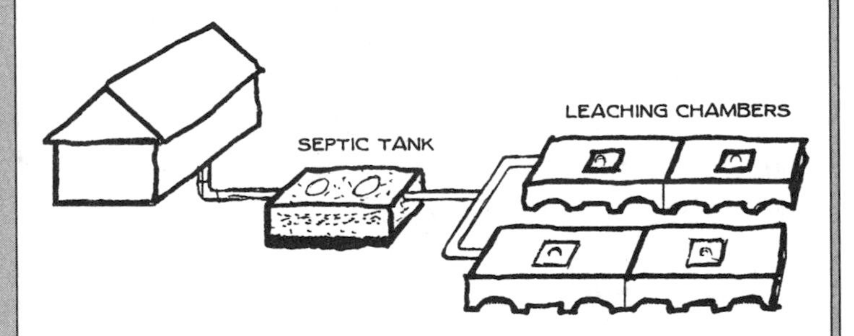

4 Septic Tank with Alternating Trenches

One set of trenches rests while the other treats the liquid from the septic tank. This design extends system life and provides a backup should one field clog. For system repairs, a new field and valve box may be added to the old system. The new field works while the old field rests and renews. Switch the fields annually in the summer.

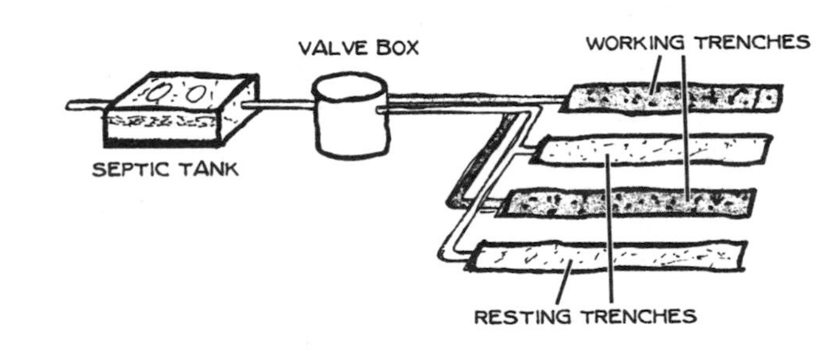

5 Pressure-Dosed Distribution

A pump or siphon doses a pressure distribution manifold that disperses the effluent evenly to each trench. Dosing prolongs system life by flooding a larger area and by forcing the exchange of air in the soil. Dosed systems are more common for larger flows. The pressure manifold can include valves or plugs that permit more control over trench loading or trench resting. Annual inspection is suggested.

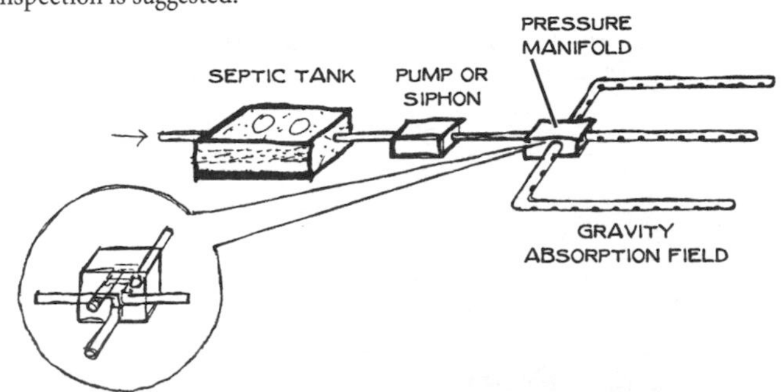

6 Shallow Trench Low-Pressure Pipe Distribution

Small-diameter pipe, located at a more shallow depth than a conventional system, receives pumped effluent. Effluent moves under pressure through small holes in the pipe and soaks the entire trench network area. Even dosing of more open and aerobic soil horizons improves treatment. Used in areas with high groundwater or shallow soil (because it places the treatment higher in the soil profile), or on steep slopes that require hand excavation. Professional maintenance is needed to flush the lines annually.

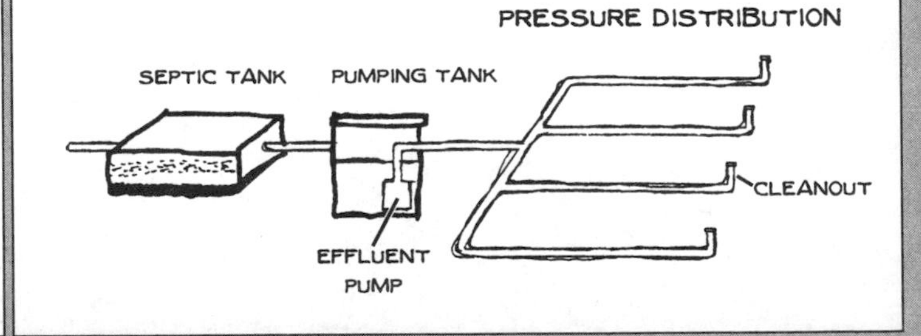

OPTIONS FOR DIFFICULT SITES

7 Pretreatment & Soil Absorption

Pretreatment addresses the need to treat higher strength waste (such as that from restaurants) and can help repair biologically overloaded systems where no additional absorption area is available. Aerobic treatment systems and filters can be used for this purpose. For aerobic treatment (called "package plants"), wastewater and air mix in a tank. Bacteria grow in the tank and break down the waste. For filters, septic tank effluent passes over porous media that trap the solids. Bacteria that grow in the media break down the waste. Professional maintenance by certified operators and a lot of energy are required for aerobic systems.

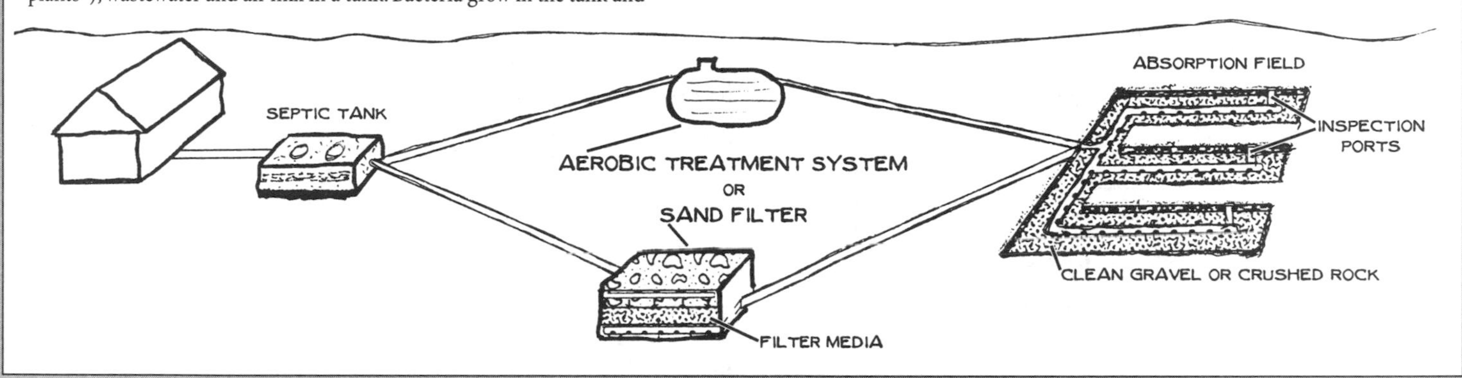

8 Septic Tank & Mound System

Pumps dose effluent into a gravel bed or trenches on top of a bed of sand. Sandy soil, carefully placed above the plowed ground surface, treats the effluent before it moves into the natural soil. The system extends onsite system use in areas with high groundwater, high bedrock, or tighter clay soils. Regular inspection of the pumps and controls and flushing of the distribution network are needed.

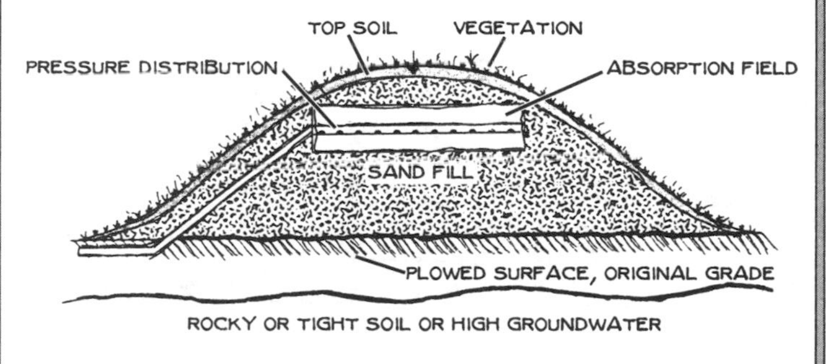

9 Evaporation & Absorption Bed

Effluent from a septic tank flows into gravel trenches or chambers in a mound of sandy soil. Less permeable soil, placed at the surface of the mound, sheds rain from the system. Trees that grow around the system and plants on top of the system pull liquid from the sand and transpire the water into the air. Some effluent may seep into the soil. This system requires a climate where evaporation consistently exceeds rainfall.

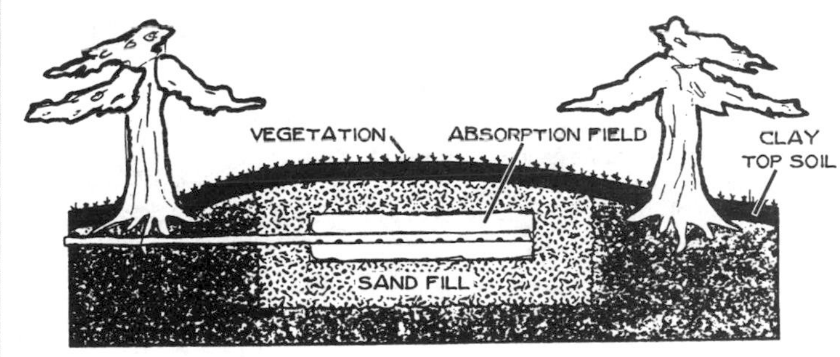

10 Septic Tank, Sand Filters, Disinfection & Discharge

Open or buried beds of sand may receive single or repeated applications of effluent. Effluent passes through the media into drains from the gravel and pipe network below the filter. Effluent may be discharged to the environment directly or into a soil absorption or land treatment system (#16). Disinfection often precedes discharge into a stream or land irrigation. Certain types of filters can significantly reduce nitrogen and may be used in areas where soil absorption is not possible. Requires inspection and periodic maintenance. Surface discharge requires management.

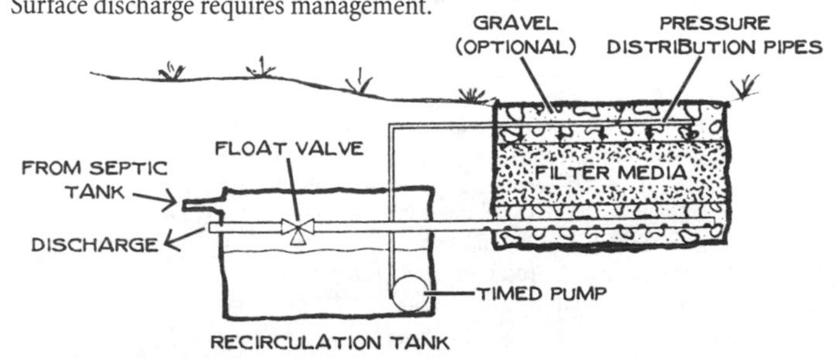

11 Constructed Wetlands

Effluent from a series of septic tanks passes through a bed of rocks planted with reeds. Liquid evaporates and drains into a soil absorption system or discharges. Used for additional treatment or where soil is not suitable for absorption. Discharge usually requires disinfection.

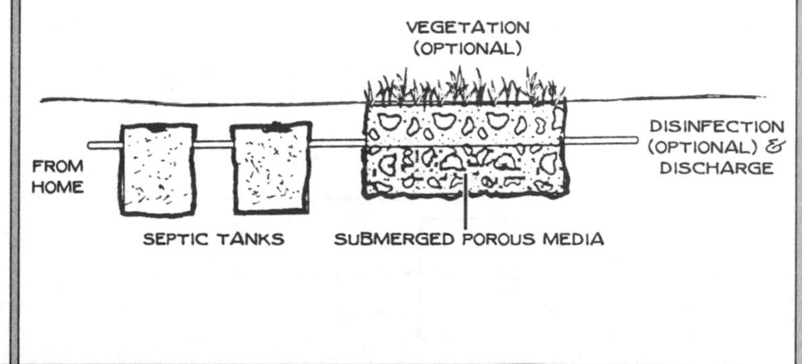

OPTIONS FOR SPECIAL SITUATIONS

12 Holding Tank

Sewage flows from low-flush toilets and water-saving fixtures into a large, watertight storage tank. An alarm in the tank signals the owner to have the sewage hauled away. Only recreational housing utilizes holding tanks because of the high hauling cost. Public management is frequently required. Contracting for hauling helps to reduce costs.

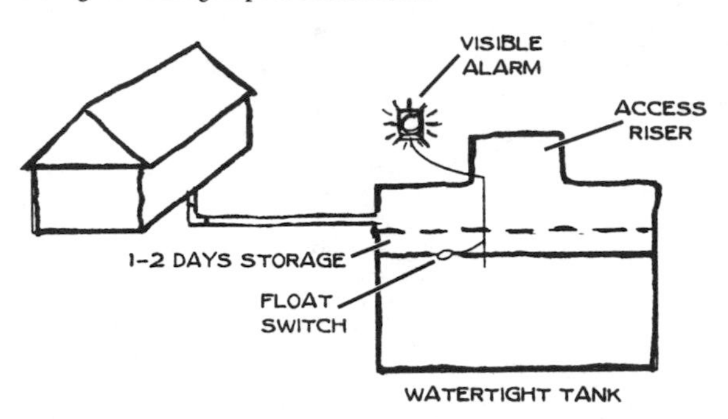

13 Lagoon

A series of septic tanks or other pretreatment systems (#7, #10, #11) discharge into a lagoon. Sunlight and long storage times support the natural breakdown of the waste and die off of harmful organisms. Effluent evaporates, slowly seeps into the soil, or receives further treatment through land application (#16). Onsite lagoons require large lots and may be fenced.

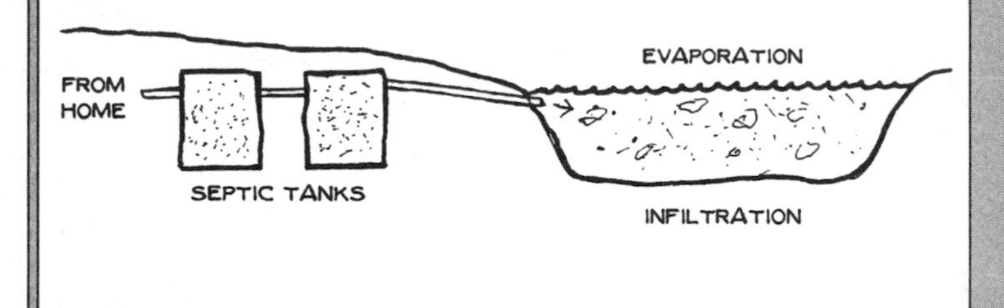

14 Waterless or Ultra Low-Flush Toilet System

Composting Toilets: No water
Serve commercial and single-family units. Well-designed units produce a dry mixture that should be managed by professionals. Reduces discharge of nutrients into water resources. Electric vent, fan, and heating element common. Proper care is essential.

Incinerating: No water
Electricity, gas, or oil burns solids and evaporates the liquid, which is vented to the roof. Small amounts of ash are removed weekly. Proper care is essential. Limited to less frequent use sites, such as recreational cabins.

Water Conservation Toilets: Low water
Low-flush toilets use 1.6 gallons or less per flush. They generally cost slightly more than conventional units, but pay for themselves by lowering the water bill. They perform well. Many work as well as 4-gallon-per-flush models.

Recycling Water: Low water
Treated wastewater or graywater recycles to flush toilets. Treatment systems use electricity and require maintenance.

15 Dual Systems

Two systems treat the waste. Composting toilets or low-flush (1.6 gallons or less) toilets coupled with a holding tank (#14, #12) exclude nutrient-rich toilet wastes (blackwater) from the wastewater disposal system. All other household wastewater (graywater) must be treated in an approved septic tank and absorption system, which is usually smaller.

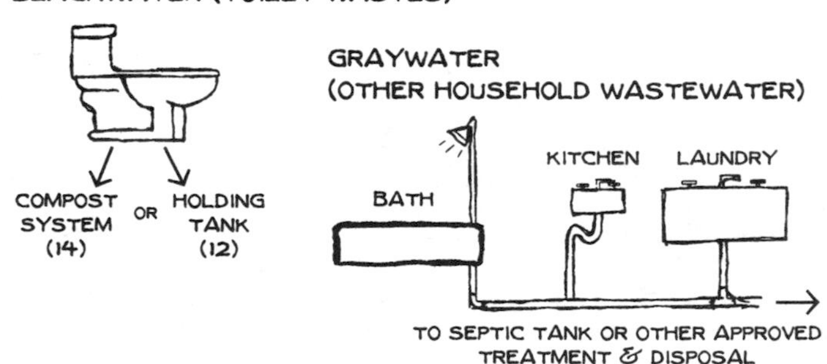

16 Land Application

Effluent from a septic tank is further treated (#7, #10, #11, #13) and stored. Timed sprinklers apply the effluent at night or below the soil surface to plants and trees in a large treatment area. Protects high groundwater in more permeable soils as plants take up nutrients and water. Disinfection and fencing may be required for individual home use. More common in warm climates, but not widely permitted by health authorities.

SLOW-RATE LAND TREATMENT

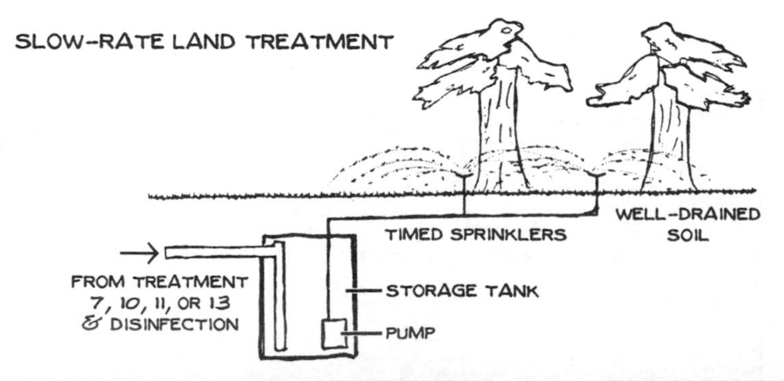

OVERLAND FLOW

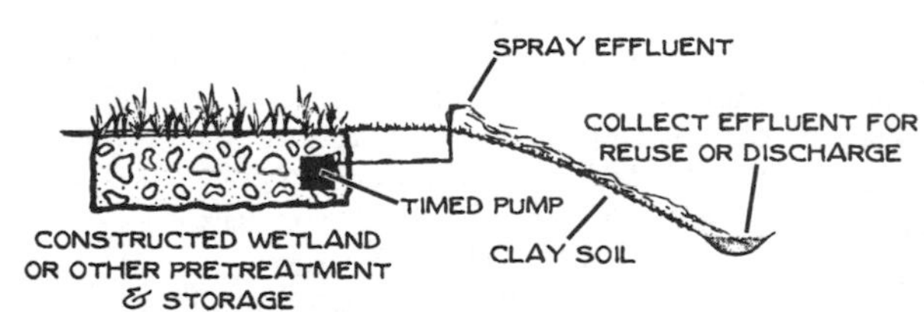

Treated effluent from a lagoon (#13) or wetlands (#11) is sprayed on the surface of a gentle, grass-covered slope. Effluent flows over the clay soil through the grass and collects at the base, where it is disinfected before being discharged. Best for tight soils where absorption systems are not possible. A professional operator usually cares for the grass and disinfection system.

COLLECTION OPTIONS FOR OFFSITE TREATMENT

17 Small-Diameter Gravity Sewers

Liquid from a septic tank flows under low pressure in 3-inch or larger collection pipes. Houses below the pipe must use small pumps (septic tank effluent pumps such as #19A and #20). Houses higher than the pipes may drain by gravity. Larger developments favor treatment by a discharging technology such as #11, #13, #16, or #19. Common in rural areas where the community treatment site is generally downhill. Central management is required.

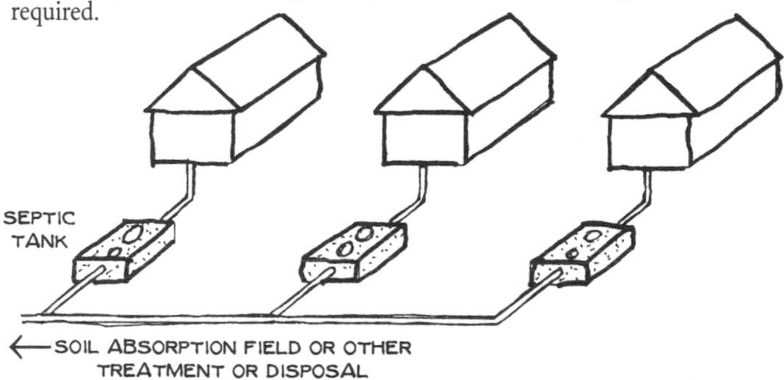

18 Vacuum Sewers

A vacuum station maintains a vacuum in the collection lines. When the sewage from one or several homes fills the storage pit, a valve opens, and the sewage and air rush into the collection line toward the vacuum station. Pumps in the vacuum station transfer the sewage to a treatment system. Power is required only at the vacuum station. Most economical where many homes are served or in areas with high excavation costs and lift stations. Requires a professional operator.

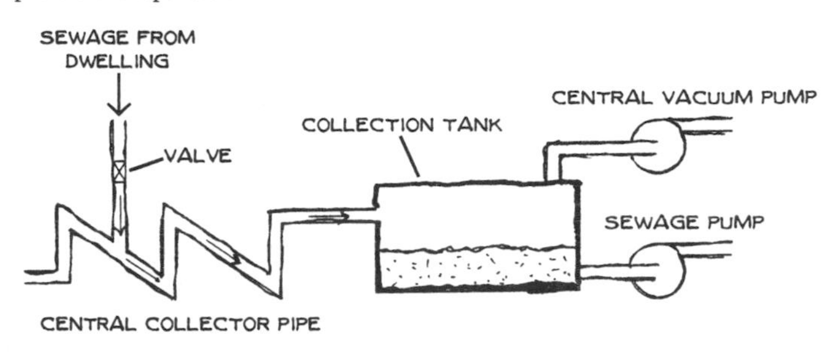

19 Pressure Sewers: Grinder Pump (GP) or Septic Tank Effluent Pump (STEP)

Sewage is first pretreated in a septic tank or grinder pump and then a pump forces the liquid through small-diameter lines to a conventional gravity sewer or to a neighborhood treatment plant such as #10, #11, #13, or #16. The community usually owns and operates shared pumping units. Plastic lines located near the surface ease installation and reduce cost. Best for low-density or slow-growth areas or where conventional sewers are costly. Central management is required.

SEPTIC TANK EFFLUENT PUMPING SYSTEM

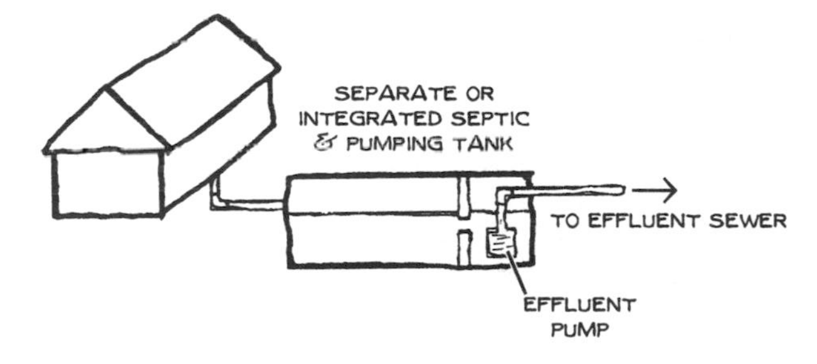

GRINDER PUMPING SYSTEM

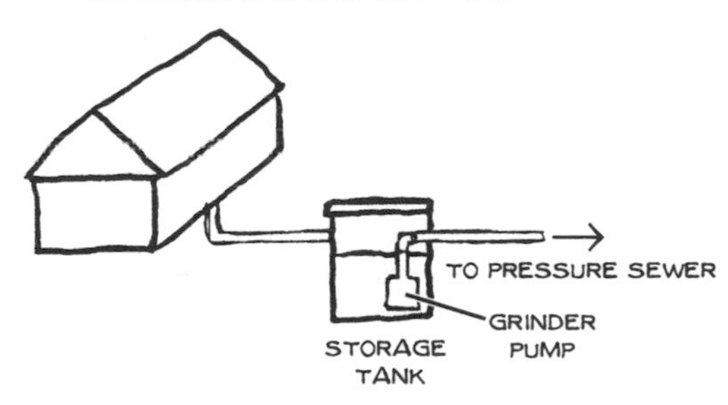

20 Alternative Effluent Collection System

Liquid from most onsite septic tanks flows by gravity in small-diameter effluent lines (#17) to a small neighborhood pump station on public property. A few homes below the sewer may also use small effluent pumps. The neighborhood lift station stores the liquid — then pumps it into a higher-pressure sewer going to a treatment system. This design can cut costs in flat terrain or where one pump unit can easily serve a number of homes. Central management is required.

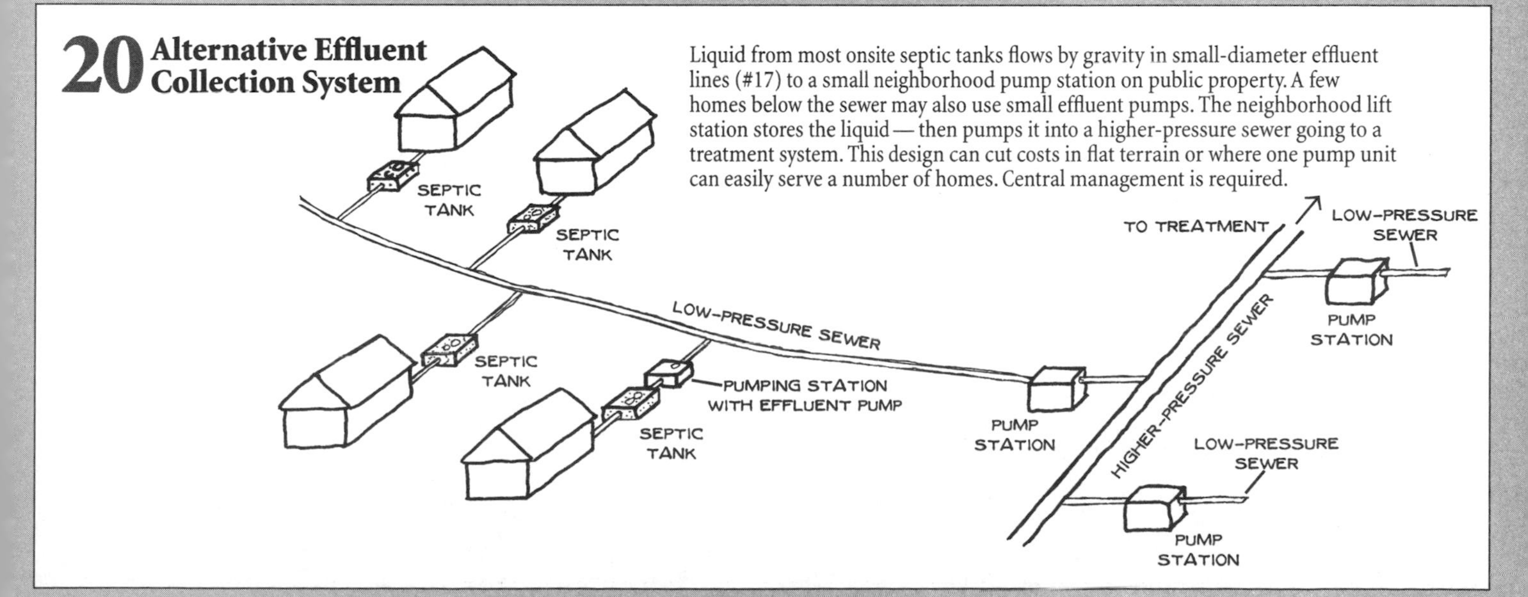

SMALL TOWN SPRINKLES LAWNS WITH WASTEWATER

"DEVELOPERS (PARKER DEVELOPMENT COMPANY AND CATELLUS RESIDENTIAL GROUP) ARE LAYING PIPES THAT WILL IRRIGATE FRONT AND BACK YARDS OF AN EL DORADO COUNTY, CALIFORNIA COMMUNITY WITH RECYCLED WATER. LOCATED IN THE DRY HILLS EAST OF SACRAMENTO, THE SERRANO DEVELOPMENT OF 4,500 HOMES ULTIMATELY WILL BE THE STATE'S SECOND AND LARGEST RESIDENTIAL DEVELOPMENT TO USE RECYCLED WATER. THE FIRST 67 LOTS TO INCLUDE CONNECTIONS TO RECYCLED WATER WENT ON SALE THIS MONTH. WASTEWATER FROM TOILETS, FAUCETS, AND OTHER ORDINARY HOUSEHOLD SOURCES WILL BE PIPED TO A TREATMENT PLANT, THEN PUMPED BACK TO HOMES IN PIPES THAT ARE DISTINCT FROM THOSE THAT CARRY POTABLE WATER. THE PIPES CARRYING RECYCLED WATER WILL BE LINKED TO FRONT AND BACK YARD SPRINKLER SYSTEMS. TO PREVENT PEOPLE FROM INADVERTENTLY DRINKING THE RECYCLED WATER, THERE WILL BE NO WAY TO ATTACH A HOSE TO THE SYSTEM. SERRANO'S USE OF RECYCLED WATER MEANS THAT THE EL DORADO IRRIGATION DISTRICT WILL DRAW LESS WATER FROM THE AMERICAN RIVER. AND WITH THE TIGHTER WASTEWATER QUALITY STANDARDS EXPECTED IN COMING YEARS FROM STATE AND FEDERAL REGULATORS, 'IT MAY BE MORE COST-EFFECTIVE TO RECYCLE WATER AND USE IT LOCALLY,' SAYS CECILIA JENSEN, WATER RECYCLING PROGRAM MANAGER FOR THE SACRAMENTO REGIONAL COUNTY SANITATION DISTRICT, 'THAN TO CONSTRUCT A BILLION-DOLLAR TREATMENT PLANT ONLY TO TREAT WATER AND THROW IT AWAY.'"

–*THE SACRAMENTO BEE*, NANCY VOGEL, 19 JUL 1999

13
A Brief History of Wastewater Disposal

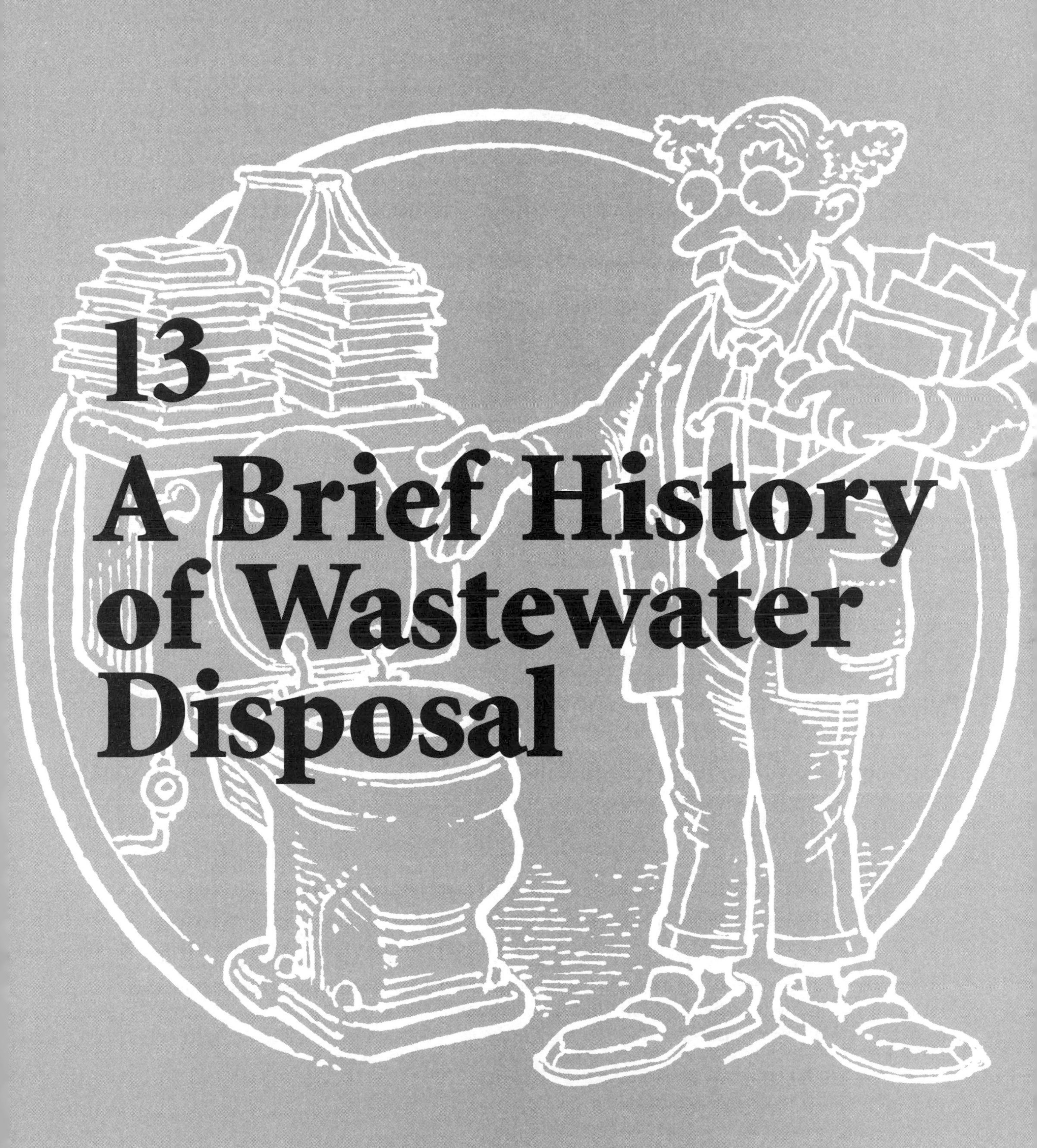

"The subject is a peculiar one."

–George Jennings, 1851

Our stone age ancestors apparently lived by the waterside: the earliest Paleolithic tools discovered were found in gravel alongside rivers. Since these people had no way of bringing water to their homes, they brought their homes to the water.

Once people moved away from the riverside (or lake, stream, waterfall, etc.), and water had to be carried, it obviously became more precious. Drinking and cooking would have come first, with bathing at home rare. The exertion of carrying water even a short distance would have outweighed the refreshing effects of bathing in it.

At some point in history, there came the revolutionary concept of channeling running water to human dwellings, and then, of using running water to carry away wastes. And with greater concentrations of populations and people settling in fixed locations, disposal of household (and later, industrial) wastes became increasingly important.

And so our short history begins.

In this book we are concerned with onsite and (in most cases) waterborne waste disposal, and in this regard, we will consider a bit of the historical background on the subject. This is by no means a comprehensive history, for that would require a book in itself. Rather, it is a collection of significant—or interesting, or amusing—practices and inventions that have led to present-day practices.

It is also admittedly Anglocentric; we cover mainly English and American history, since all of our references were in the English language. However, it is worth noting that England was at the forefront of European sanitation inventions of the last 400 years.

NEOLITHIC DRAINAGE

Waterborne waste disposal dates back at least 5000 years, when Neolithic people of the Orkney Islands (off the coast of Scotland) constructed drainage channels designed to carry wastes from their homes into a nearby body of water. A number of dwellings at what is now called the Barnhouse Settlement, near the Stones of Stenness in the Orkneys, had stone-built drainage channels running from recesses in the walls, which were supposedly latrines. The drains led to the nearby Loch Harray. The settlement was occupied around 3500 B.C.

THE FIRST FLUSH TOILETS

In the ancient cities of the Indus Valley, now Pakistan, from about 2500 to 1500 B.C., some houses had bathrooms with water-flushed toilets. Wastewater was channeled into street drains via brick-lined pits, with outlets about three-quarters of the way up, much like today's septic tanks.

In 1700 B.C., plumbing was in a very advanced state on the island of Crete in the Mediterranean Sea. The palace of Minos, in the town of Knossos, was a complex architectural achievement, with hundreds of rooms built around a central courtyard. It featured bathtubs filled and emptied by terra-cotta pipes fitted together very much like present-day plumbing *(see next page)* and—amazingly enough—flush toilets with wooden seats and overhead reservoirs for flushing. Excavations reveal four separate drainage systems that emptied into large sewers built of stone.

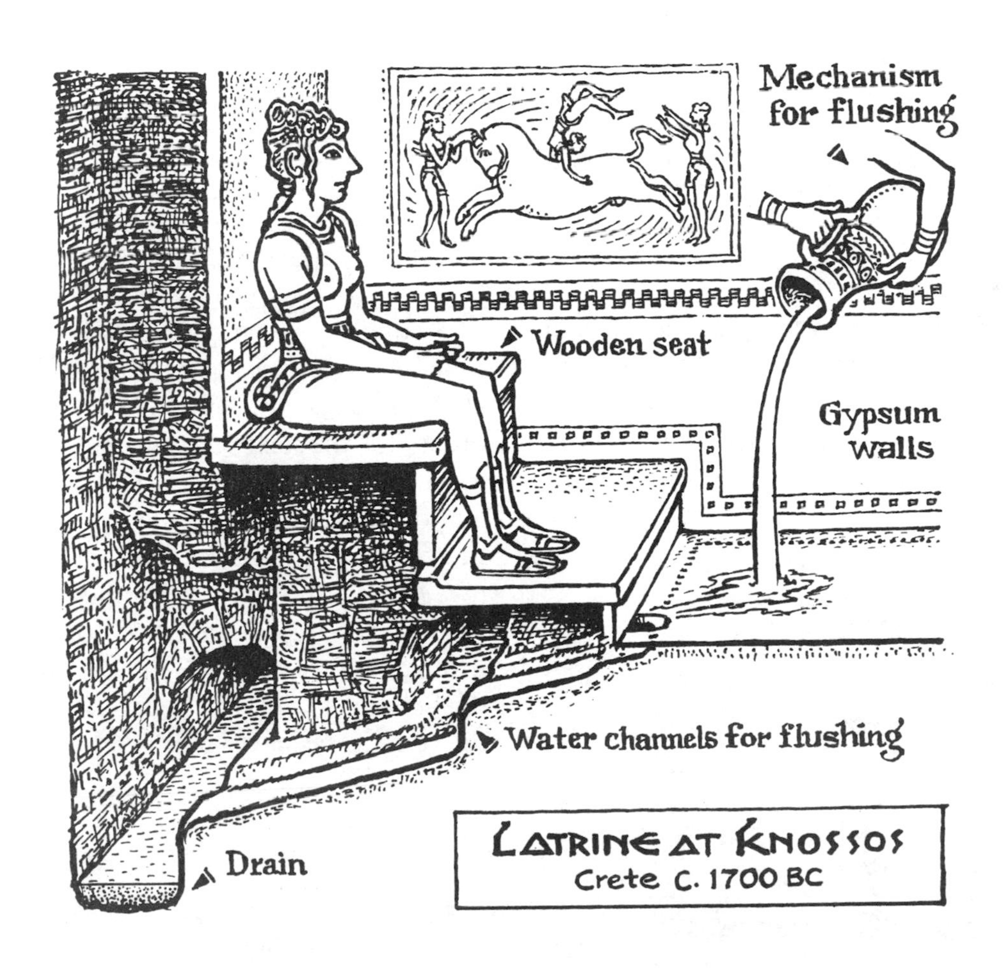

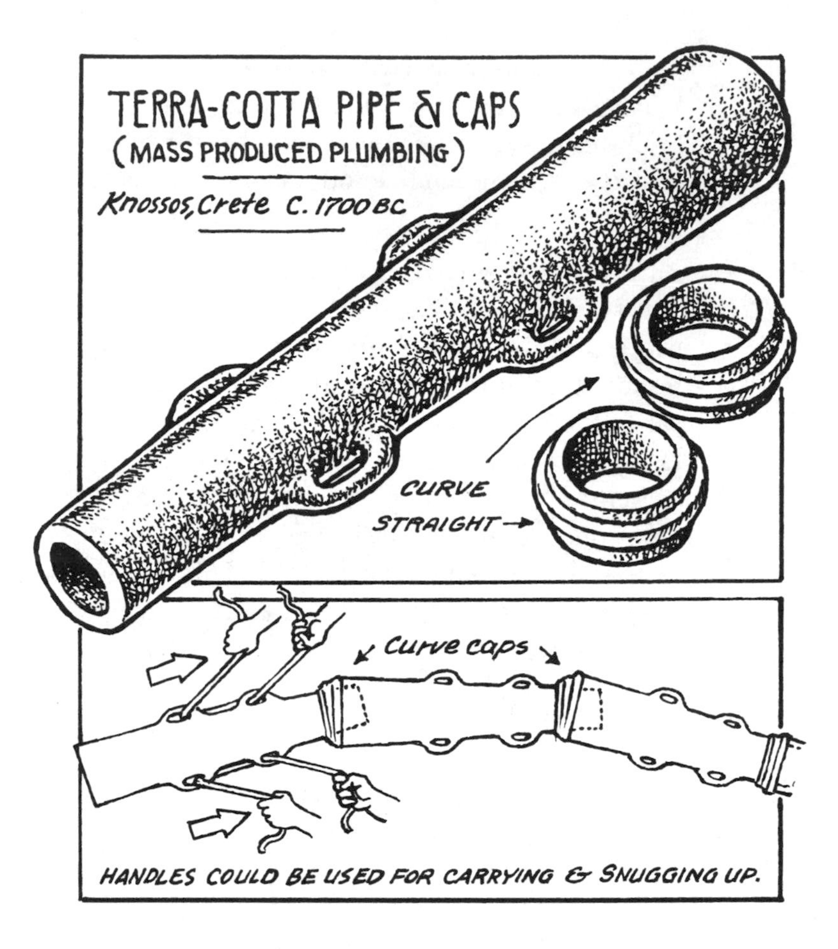

Terra-cotta pipes excavated in Knossos, Crete, circa 1700 B.C., are similar to pipes today.

A small bathroom excavated in the city of Akhnaten at Tel-El-Amarna in 14th-century Egypt featured a keyhole-shaped opening in a limestone seat. The waste dropped into a removable vase, in a pit below. No water was used for flushing.

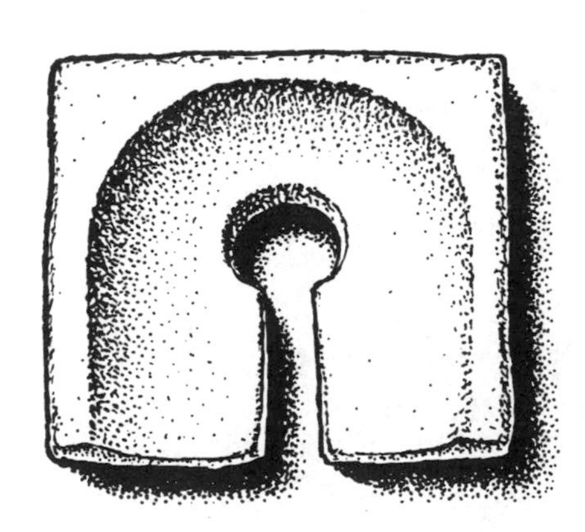

Limestone seat from 14th century Egypt
Waste fell into a removable container below.

The convenience and practicality of the flush toilet that was enjoyed by Minoan royalty some 4000 years ago was apparently available to few others for the next 35 centuries. For it was not until 1596 A.D. that Queen Elizabeth I's godson, Sir John Harrington, designed and built what he called "a privie in perfection" for the queen. *(See p. 144.)* In the intervening years, there are no records of water-flushed toilets. Although Leonardo da Vinci (1452–1519) designed a water-flushed toilet with pipes venting gasses through the roof, it was never built.

ROMAN PLUMBERS AND THE ART OF BATHING

Two of the great physical achievements of the Roman Empire were its system of roads and its highly developed art of plumbing. *Plumbus* means "lead" in Latin, and Roman plumbers (many of whom were women) plumbed and soldered pipes for hot and cold water systems in homes as well as in gigantic bath houses.

The Baths of Caracalla in Rome covered a 27-acre site. Series of aqueducts were built to transport water to Rome; in 52 A.D., there were 220 miles of Roman aqueducts, 30 of them above ground. At its peak development, the system carried some 300 gallons per person per day to Rome.

At Aqua Sulis, the site of an ancient spa and hot springs in England (the present-day Bath), the Roman Emperor Claudius built a magnificent bathing complex in 43 A.D. The lead pipes laid down some 2000 years ago for hot water are still functioning. By the 4th century A.D., Rome had 11 public baths, over 1300 public fountains, and 856 private baths. Not only were there private water-flushed toilets, there were public ones. In 315 A.D., Rome had 144. Some homes in Pompeii had 30 water faucets.

By 500 A.D., when the Roman Empire was in its decline, invading barbarians swept across Europe, destroying most of the tiled baths and water systems. As Lawrence Wright says in his book, *Clean and Decent,* "the taps were being turned off all over Europe; they would not be turned on again for nearly a thousand years." Sanitation technology entered its dark ages.

Ruins of latrines in Hadrianic Baths in the Roman town of Lepcis Magna, North Africa, 1st century A.D. *The seats extended around three sides of the room and it was, no doubt, a busy social center during peak hours. This system spared no water. Aqueducts brought in large amounts of water, which ran to hundreds of fountains and replenished the baths, thence under the latrines and streets and out to sea. There were no shut-off valves.*

THE AGE OF DISEASE

Many early Christians rejected most things Roman, including the value of cleanliness. The body and things of the flesh, including bathing, were considered evil by many church leaders. "To those that are well, and especially for the young, bathing shall seldom be permitted," said St. Benedict. Being unwashed was apparently considered holy by St. Francis of Assisi.

With the demise of private as well as public bathing went the conveniences of indoor plumbing. The outhouse, open trenches, and the chamber pot reemerged at all levels of society. Christian prudery put a damper on sanitation throughout most of Europe.

However, it must be pointed out that in those days when "for a thousand years Europe went unwashed," there were some monasteries that continued the art of plumbing and bathing. Many abbeys in Britain had piped water before 1200. The Christchurch Monastery at Canterbury, for example, had running water, purifying tanks, and wastewater drainage from toilets, and the monastery was spared from the Black Plague in 1349.

NOT-SO-BUCOLIC VILLAGES

In early rural life, a typical peasant family lived in a one-room hut with a dirt floor. Smoke from the fire vented through the thatched roof. The floor, often covered with straw, was a haven for lice and vermin. People often relieved themselves in a corner of the room or in the muck and mire outside.

In medieval towns, the idealized version of Merrie Olde England may have been fanciful retrospect. In *Health, Wealth, and Population in the Early Days of the Industrial Revolution*, M. C. Buer rejects the notion that unhealthy conditions were brought on solely by the advent of the industrial revolution. Habits that might be harmless in a rural setting could prove fatal in a town, where sanitary regulations were seldom enforced.

GARDEROBES IN CASTLES

In European castles, with thick stone walls, openings like cupboards were built into the castle walls high up near or on the battlements. There a seat with a hole in it and the chute below it led diagonally out through the wall so excrement fell into the moat or river or pit below. This device was called a garderobe, a euphemism for the medieval privy.

Straight-on view

THE STENCH IN CITIES

Things were even more grim in the cities. London's sewers were open ditches in the middle of the street that drained into the Thames. Cesspits (closed receptacles for excreta and refuse) were built under houses and their contents had to be removed by hand; they overflowed into the ditches when full. Cesspit wastes often soaked foundations, walls, and floors of living quarters. (A cesspit differs from a cesspool in that the latter is designed to allow liquids to escape into the surrounding soil.) Cesspits were supposed to be leakproof, but they generally leaked, often contaminating drinking water from wells.

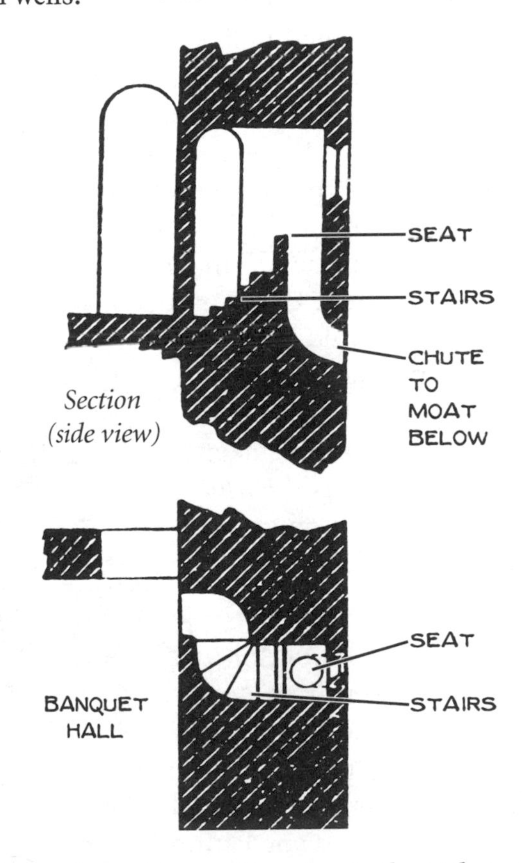

Garderobe in the Tower of London in the 11th century. *A small recess built into the three-foot-thick stone wall of the banquet room, with a narrow window. A short curved shaft discharged down the face of the outside castle wall into the moat. As stated in the book* Clean and Decent, *"Formerly defensive, moats must have become offensive."*

Chamber pots were popular in the 1400s, and by the 1800s were highly developed. Some, made of pottery, became a "sure source of naughty fun," such as those with a large eye in the target area with the lines:

> *Use me well and keep me clean*
> *And I'll not tell what I've seen.*

Some had bottom portraits of national villains, such as Napoleon, and, much later, Adolf Hitler. Some were rigged to play jokes. One 19th-century pot had a hidden music box that played when the pot was lifted and could not be turned off.

In tenements, wastes were stored in chamber pots and then thrown out the doors or windows into the streets. The English slang for toilet, *loo,* is said to come from the expression, "Gardy-loo" (in French *Gardez l'eau*), meaning "Watch out for the water!" In Edinburgh, excrement from the previous 24 hours was thrown into the street, often from as high as 10 stories, and it sat there until the early morning when it was cleared away by the City Guard. On Sundays, it could not be touched, and sat there all day, "filling Scotland's capital with the savour of a mistaken piety." (*English Social History,* G. M. Trevelyan, Longmans, Green and Co., London, 1944.)

The Thames in London was an open sewer. It was the same in Paris. A French magazine of the 1700s proclaimed, "Paris is dreadful. The streets smell so bad you cannot go out.... The multitude of people in the street produces a stench so detestable that it cannot be endured."

THE ANGEL OF DEATH

The Black Plague, which started in the 1300s, killed an estimated one-third of Europe's population. It was caused by fleas and ticks that lived on host black rats, which fed off the readily available garbage and excrement. Other diseases directly related to human waste wiped out many hundreds of thousands of people in the Middle Ages. These included dysentery, typhus (which comes from bad sanitation and is highly contagious), and typhoid fever (from human feces and urine).

Cholera, another waterborne disease which comes from human feces and can cause death within a few days, became the most feared disease of the 19th century. A worldwide epidemic of cholera was precipitated by the crowded living conditions during the Industrial Revolution in England and the United States and raged as well throughout other parts of the world, including Europe, Russia, India, and Iran.

JOSÉ GUADALUPE POSADA

THE AGE OF SANITATION

In the mid-1800s, Dr. John Snow demonstrated that cholera in London was waterborne and taken into the system orally. It wasn't until 1876 that Dr. Robert Koch of Germany discovered the comma-shaped bacillus of cholera under a microscope; he determined that excrement could contain the infectious cholera bacteria for some time. Up to this time in history, people did not understand that contagious disease could be transmitted by unseen germs in contaminated water and via unsanitary waste disposal. They didn't believe something invisible to the naked eye could cause illness. The general ignorance about sanitation gave way as microbiologists discovered that sanitation and a certain level of cleanliness were essential to prevention of disease.

Cholera was worst in the crowded slums of cities, and for a while it seemed to the rich a disease of the "unwashed masses." To the poor it at times seemed like a plot against them. When cholera first hit Paris, there were so few deaths outside the lower classes that the poor regarded the epidemic as a plot hatched by the rich to get rid of them.

The immunity of the rich was, however, short-lived. The open sewers in the cities eventually contaminated wells and/or ran into waterways that supplied drinking water to all. Once the rich and privileged started dying, government reform got underway.

After Snow's discovery, English authorities began campaigning for better sanitation in the home, workplace, and public streets. British engineers led the way in sewer construction and separation of wastes from drinking water. The English Public Health Code of 1848 became a model plumbing code for the rest of the world. Yet it would be several decades before the improvements would wipe out cholera and typhoid.

By 1830, there were 2 million people living in London. Chamberpots were emptied into the streets and raw sewage ran into the Thames. In 1849, 55,000 people in London died of cholera.

A story is told about Queen Victoria touring Cambridge University in the mid-1800s, where the sewer discharged straight into the River Cam. Turning to her guide, Dr. Whewell, Master of Trinity, she asked about all the pieces of paper floating in the river. "Those, ma'am," he coolly replied, "are notices that bathing is forbidden."

England Victorious, *a drawing by German satirist Heinrich Kley (1863–1945) showing the privy and chamber pot being defeated by Triumphant English Plumbing, which is unfurling a toilet paper banner stating, "Britannia rule* (sic) *the waves." (Brittania appears to be blushing in the background.) Indeed the late Victorian era was a time when scores of people changed over from privies and chamber pots to waterborne waste disposal.*

In an 1834 print by Hokusai, a great samurai warrior is shown squatting in a privy while his faithful retainers wait—and suffer—outside. (This appears to be a public facility since the sign asks users to keep it clean, and there is some graffiti on the wall.) The travellers bear luggage and the samurai's sandal holder and are armed to the teeth. Japan, with a fishing economy, and limited farm land, has traditionally utilized agricultural composting rather than the western flush method for waste disposal.

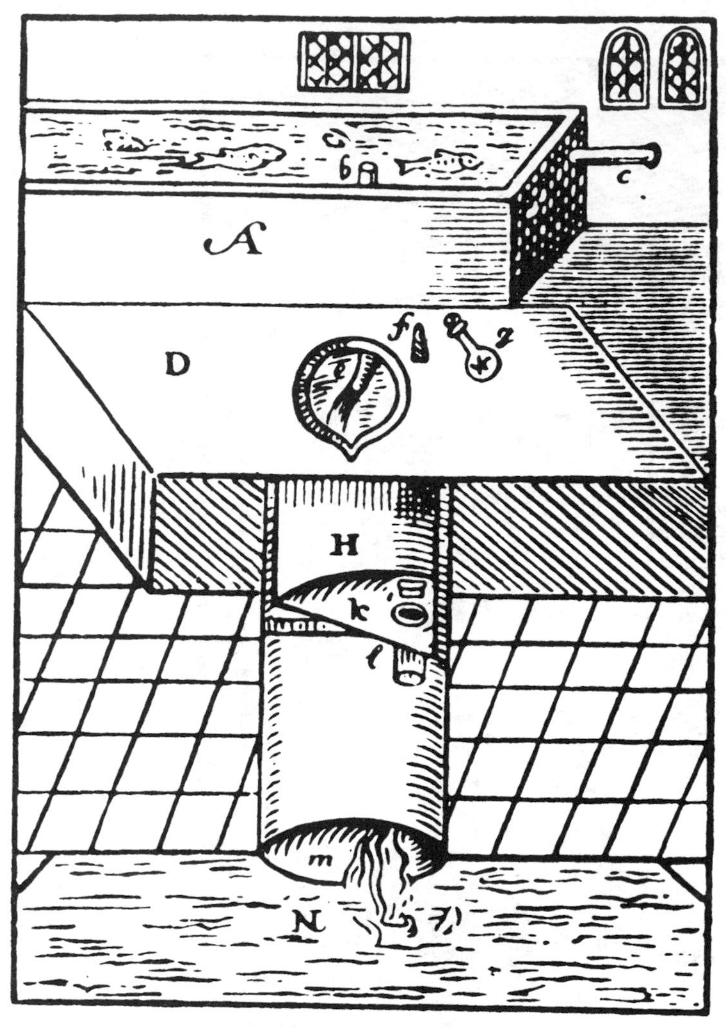

A priuie in perfection

A. the Cesterne.
B. the little washer.
C. the wast pipe.
D. the seate boord.
E. the pipe that comes from the Cesterne.
F. the Screw.
G. the Scallop shell to couer it when it is shut downe.
H. the stoole pot.
I. the stopple.
K. the current.
L. the sluce.
M.N. the vault into which it falles: alwayes remember that () at noone and at night, emptie it, and leaue it halfe a foote deepe in fayre water. And this being well done, and orderly kept, your worst priuie may be as sweet as your best chamber. But to conclude all this is a few wordes, it is but a standing close stoole easilie emptyed.

And by like reason (other formes and proportions obserued) all other places of your house may be kept sweet.

John Harrington's "Privie in Perfection," built for Queen Elizabeth's palace in Richmond in 1596, was a landmark in waste disposal. Yet it was not until 200 years after Harrington's invention that the water closet was rediscovered and eventually utilized by large segments of the population.

THE "PRIVIE IN PERFECTION"

John Harrington's water closet for Queen Elizabeth I in 1596 was as sophisticated as it was prescient. It had a tank above the bowl, a valve that could be turned on to let water flow into the tank, and a valve that discharged sewage into a cesspool below. Harrington intended a water seal, recommending that the drainage pipe be left, "after it is voided, half a foot deep in clean water." However, this was a straight pipe (not a trap [*see next page*]), and it apparently didn't work all that well. Elizabeth complained that the fumes from the sewage below discouraged her from using the device.

THE AMAZING PLUMBING TRAP

Several hundred years passed before similar water closets were reinvented in England, and then a significant breakthrough occurred. With units designed by Alexander Cummings and others in the 1770s and later, the drainage pipe under the bowl curved upward, forming an "s" shape (an "s" on its side) so that a quantity of water formed a perfect seal, blocking the entrance of foul odors from the cesspool or sewer below.

A water seal is used to this day in all plumbing drains; it is built into toilet bowls, and visible as the "p-trap" underneath all sinks. (Take a look under one of the sinks or basins in your house.) The lower portion of the curve is always filled with water, forming a watertight seal that, along with a vent pipe to the roof, keeps sewer gasses from backing up into living quarters. This brilliant invention of some 200 years ago, its principle little-known to most people even today, performs an essential role in modern plumbing and sanitation.

JOSEPH BRAMAH

Joseph Bramah improved on Cummings' closet in 1778. When one pulled the handle on Bramah's unit, it opened the valve, flushed out the excrement, and turned on water to flush the bowl. When one pushed the handle, it shut the valve and ran water into the bowl for 15 seconds, in order to be ready for the next user. Bramah's units tended to leak a bit (but not as much as Cummings' units), and they were beautifully made: a mahogany case covered the apparatus. He claimed to have sold 6,000 units, and for 100 years, they were the best in England.

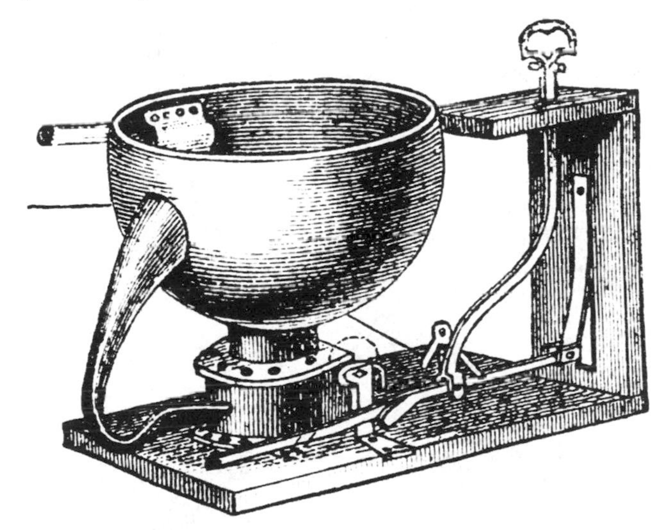

Bramah's Valve Closet of 1778 was an improvement on Cummings' earlier invention. The valve closing the outlet of the bowl was hinged instead of sliding. It was less likely to become encrusted and freeze up, so it leaked less.

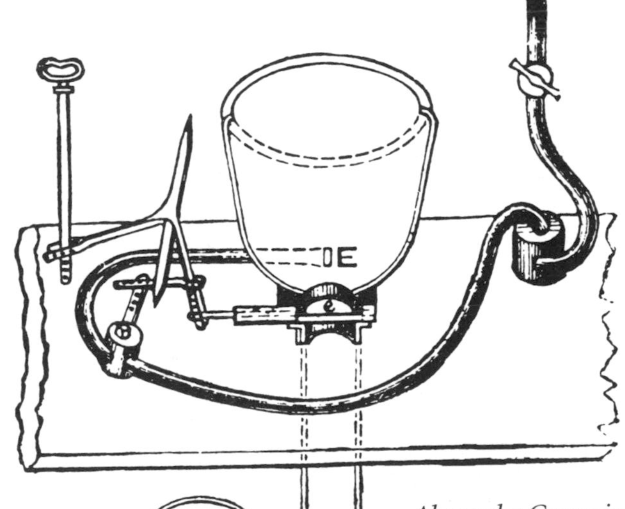

Alexander Cummings' Valve Closet of 1775, showing the revolutionary syphon trap filled with water that prevented gasses from backing up into the house. The one problem with Cummings' design was that the valve (the one closing the bowl's outlet) was unreliable.

THOMAS CRAPPER

No discussion of mid-19th-century water closet devices would be complete without mention of Thomas Crapper. Crapper was a plumber who held three patents for water closets and founded a plumbing shop that supplied a variety of bathroom products in England. The most famous product attributed to him was *The Silent Valveless Water Waste Preventer,* which allowed a toilet to flush effectively when the bowl was only half full. This saved a great deal of water, as previously people merely opened a valve and left it open for a long time. It is believed that Crapper did not invent this device, but most likely bought the rights from the inventor, Albert Giblin, and perfected it. While he was in business, Crapper conducted stringent tests on cisterns, bowls, and other toilet paraphernalia and undoubtedly made a great contribution to flush-toilet technology.

Does the slang word crap come from his name? Perhaps, but it's also been conjectured that it could come from the Dutch *krappe* or the French *crape*, meaning rejected matter, residue, or dregs. However, the word crapper, as it refers to a toilet, was apparently the result of American soldiers passing through England in World War I and seeing the words "T. Crapper–Chelsea" printed on toilet tanks. They dubbed the toilets "crappers."

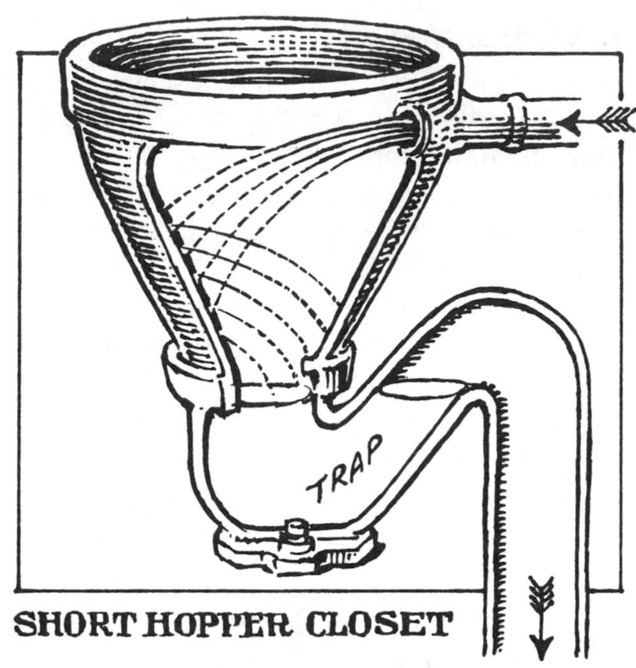

The Short Hopper Closet was a rival to Bramah's units in the mid-1800s. It featured a trap, with water spiraling in from the top of the bowl; this thin trickle of water did little to dislodge the stuff adhering to the slanted sides of the bowl.

Insignia of Thomas Crapper, Manufacturing Sanitary Engineers

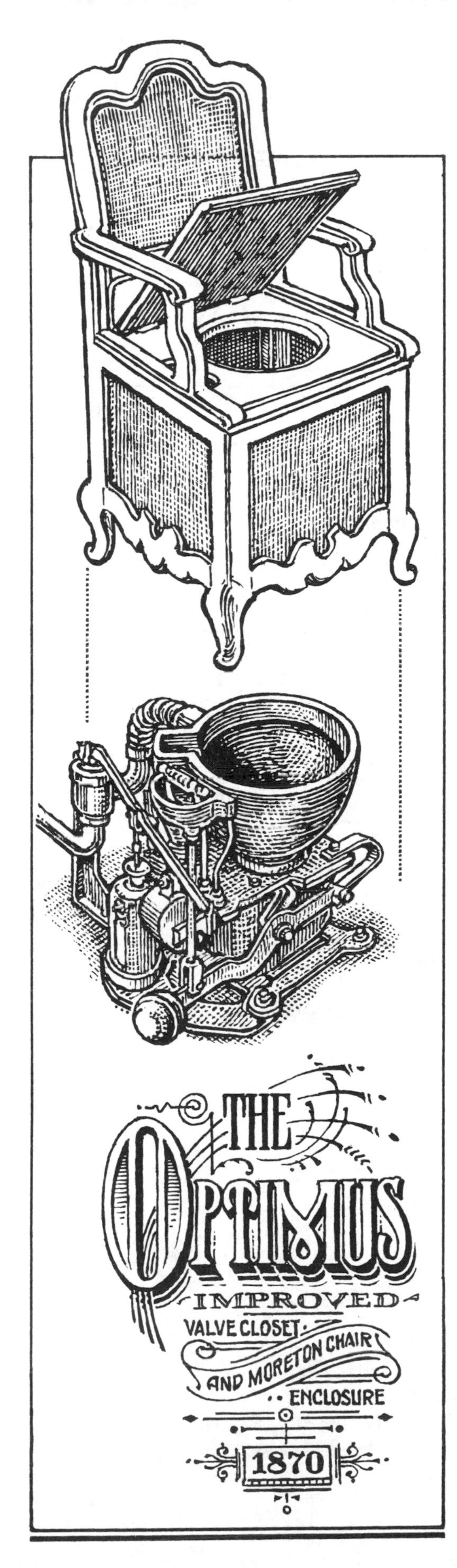

GEORGE JENNINGS

In 1858, George Jennings began a crusade in London for "conveniences suited to this advanced stage of civilisation." By 1884, Jennings' "Pedestal Vase" won the Gold Medal Award at the Health Exhibition of that year. It was acclaimed "as perfect a sanitary closet as can be made." It reportedly was able to flush:

10 apples
1 flat sponge
4 pieces of paper laid flat

with one two-gallon flush. It is said that a Mr. Shanks (still unimpressed) grabbed the hat off an apprentice's head, tossed it in, and pulled the chain, whereupon it disappeared and he cried, "It works!"

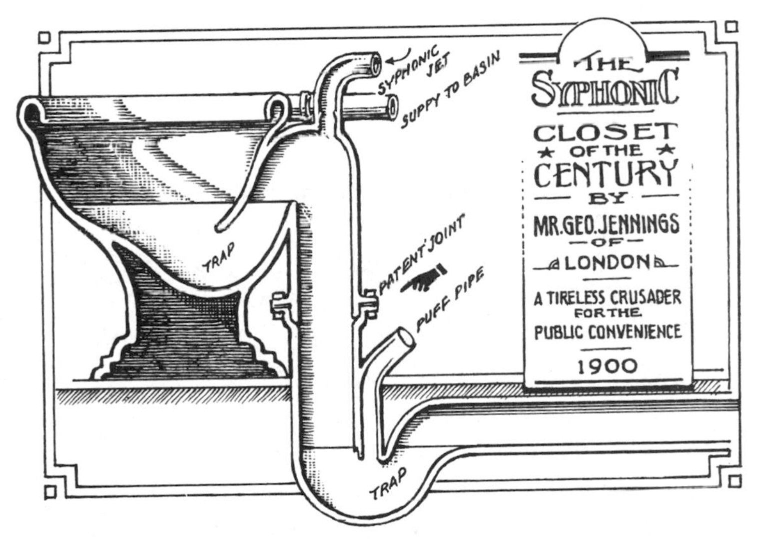

Jennings' Syphonic Closet of 1900
This unit featured standing water in the bowl and a fast flush followed by a slower second flush. Syphonic action of the water produced a clean flush. It won the Gold Medal Award at the London Health Exhibition of 1884, and was dubbed "as perfect a sanitary closet as can be made."

Stevens Hellyer's Valve Closet
Despite its mechanical complexity, the Optimus Valve Closet of 1870 by Stevens Hellyer apparently worked—many are still operating today. It flushed better (and quieter) than previous valve closets, and its seal was tight.

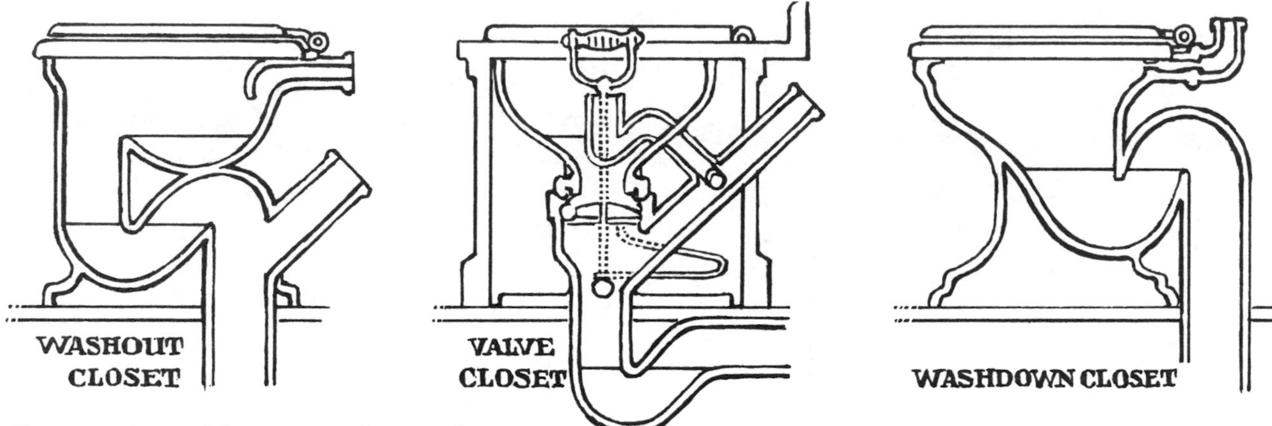

Cutaway views of three types of water closets
With the Washout Closet, the bowl held an inch or so of water and tended to flush its contents rather weakly. The Valve Closet, such as Bramah's or the Optimus, utilized a valve closing the outlet to the bowl; the Washdown Closet, as in Jennings' design, had a good quantity of standing water in the bowl and flushed its contents thoroughly.

THE GOOD EARTH

An early critic of waterborne waste disposal was Henry Moule, born in Wiltshire, England in 1801. In 1861 he produced a 20-page pamphlet entitled *National Health and Wealth Instead of the Disease, Nuisance, Expense, and Waste Caused by Cess-pools and Water-drainage.* "The cess-pool and privy vault are simply an unnatural abomination," he said, and "the water closet has only increased those evils." Moule would be in tune today with the engineers and ecologically-minded folks who criticize the use of clean water for waste disposal.

Moule believed in covering excrement with dry earth; it was sanitary, it removed odors, and the result could be used for valuable fertilizer. He designed the Moule Earth Closet, which had a receptacle chamber, back of which was a container of earth. The user used a small trowel to cover the excrement with earth.

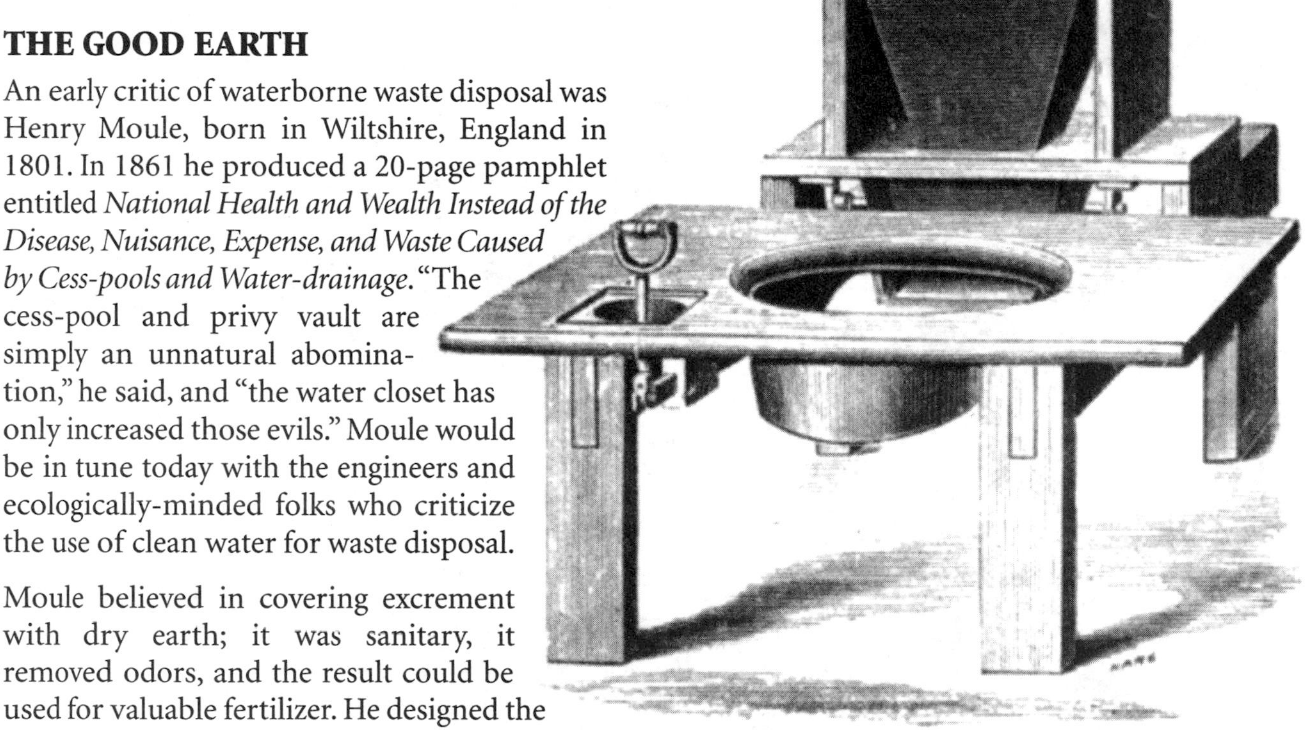

Henry Moule's Earth Closet of 1860
This unit had a wooden seat, a bucket beneath, and a hopper above and at the rear filled with fine dry earth or ashes. Pulling the handle released a layer of earth into the bucket.

No history of waste disposal would be complete without a mention of the American outhouse. Often built on a light frame so it could be moved to a new hole periodically, outhouses worked well (and still do) in rural areas with sparse rainfall. Most people don't know that the crescent moon outhouse door vent was actually used to signify the ladies' convenience. Other symbols, or none, indicated the men's facility. None of this was seen as charming in the teeming cities, where the outdoor privy had become, more or less, a plague.

THE MODERN BATHROOM

In 1868, the New York Metropolitan Board of Health was formed, the first such in the United States. The idea of sanitary plumbing systems inside buildings was an American development that soon spread throughout the world. The modern bathroom, as we know it today, began to emerge, with its central feature being the flush toilet.

In the 1870s, America was still importing English water closets but by century's end, U.S. manufacturers had caught up and were producing their own high-quality products.

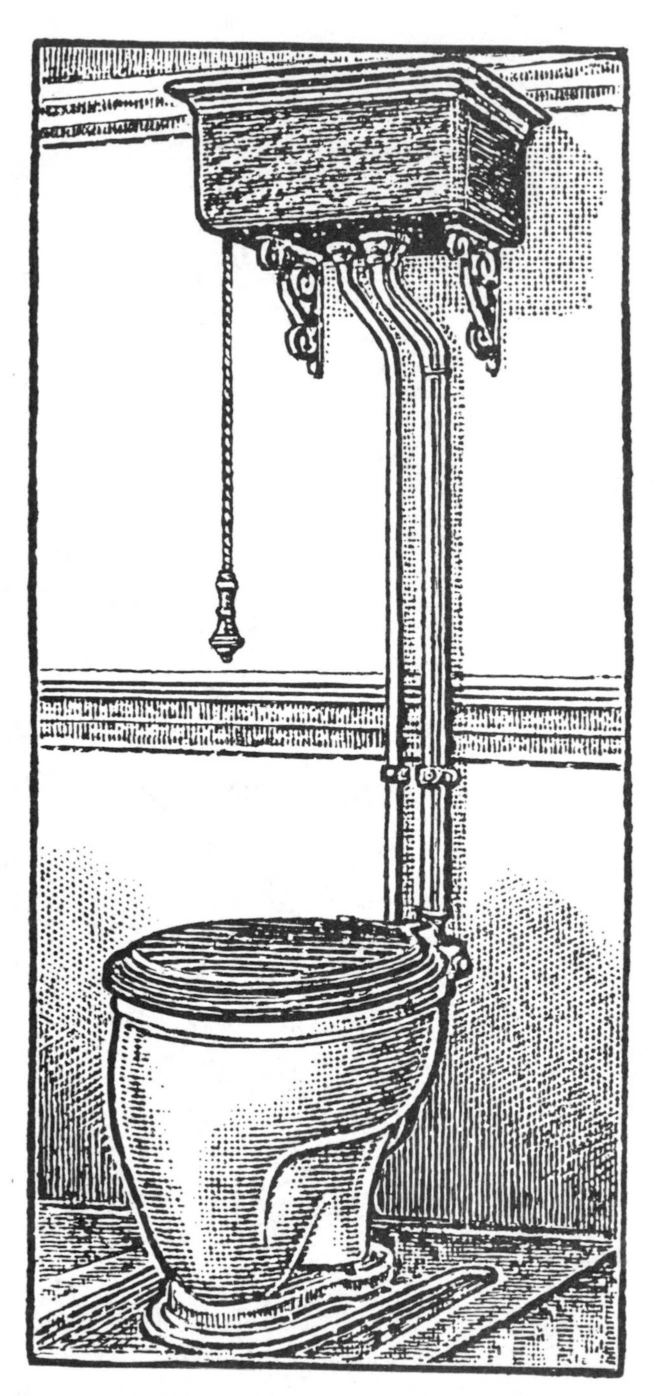

The Acme Water Closet (1902)
Oak seat and copper-lined oak syphon tank. Nickel-plated pull chain, pipes, and fixtures. No. 3 front wash-out bowl. Furnished complete. Price: $13.50.

HISTORICAL CONTEXT

What does all this have to do with the subject of this book? What do Roman aqueducts, the Black Plague, and sanitation advances in the early 19th century have to do with septic systems? The one factor they have in common is the use of *water* as a medium for transporting wastes.

The lessons of history are instructive in understanding the art of modern wastewater technology. History also teaches us the dangers of poor sanitation and the value of sensible treatment of human waste. It's worth knowing about the discoveries, the inventions, and the hardware that have led much of the world to higher levels of sanitation and health. We've come a long way from the days of rampant disease!

THE FIRST SEPTIC TANKS

Up to now in this chapter we have looked at wastewater *devices*. Now we'll take a brief look at wastewater *systems*, more particularly, septic systems, the onsite waste disposal systems utilized by some 28 million homes in America.

The septic tank was invented in 1860 by Louis M. Mouras of Vesoul, France. It was called the *Fosse-Mouras, la vidangeuse automatique,* or in English, the Mouras Automatic Scavenger. The earliest tanks were used to treat wastes primarily for community sewage. The tanks were often quite large, with capacities of 100,000 gallons or more. A tank in Birmingham, England covered over five acres.

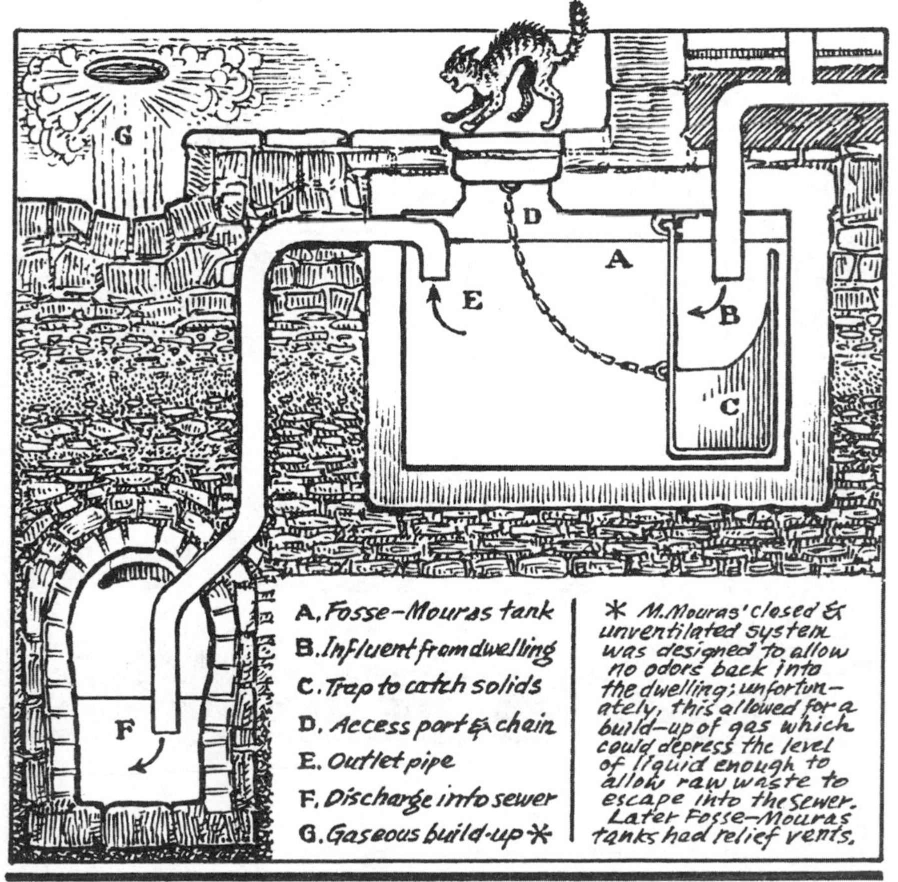

The main purpose of these tanks was to remove gross solids before discharge into the nearest stream or river. A secondary purpose was to reduce the overall volume of sludge by partial liquefaction of the solids due to digestion in the tank. With respect to these two goals, the tanks worked fairly well. Solids were retained and sludge was reduced before discharge. Yet one problem remained: effluent was largely untreated and caused pollution of streams and rivers.

In 1895, Donald Cameron of Exeter, England, gave the septic tank its name. In the 1910 text, *Sewage Disposal,* by Kinnicutt, Winslow, and Pratt, it is noted that "in 1895 he (Cameron) installed a water-tight covered basin for treatment of sewage of a portion of the city by anaerobic putrefaction and gave it the picturesque name of the septic tank, by which it has since been known." This was a 53,000-gallon tank that was 65 feet long, 19 feet wide, and had an average depth of seven feet. *(See next page.)*

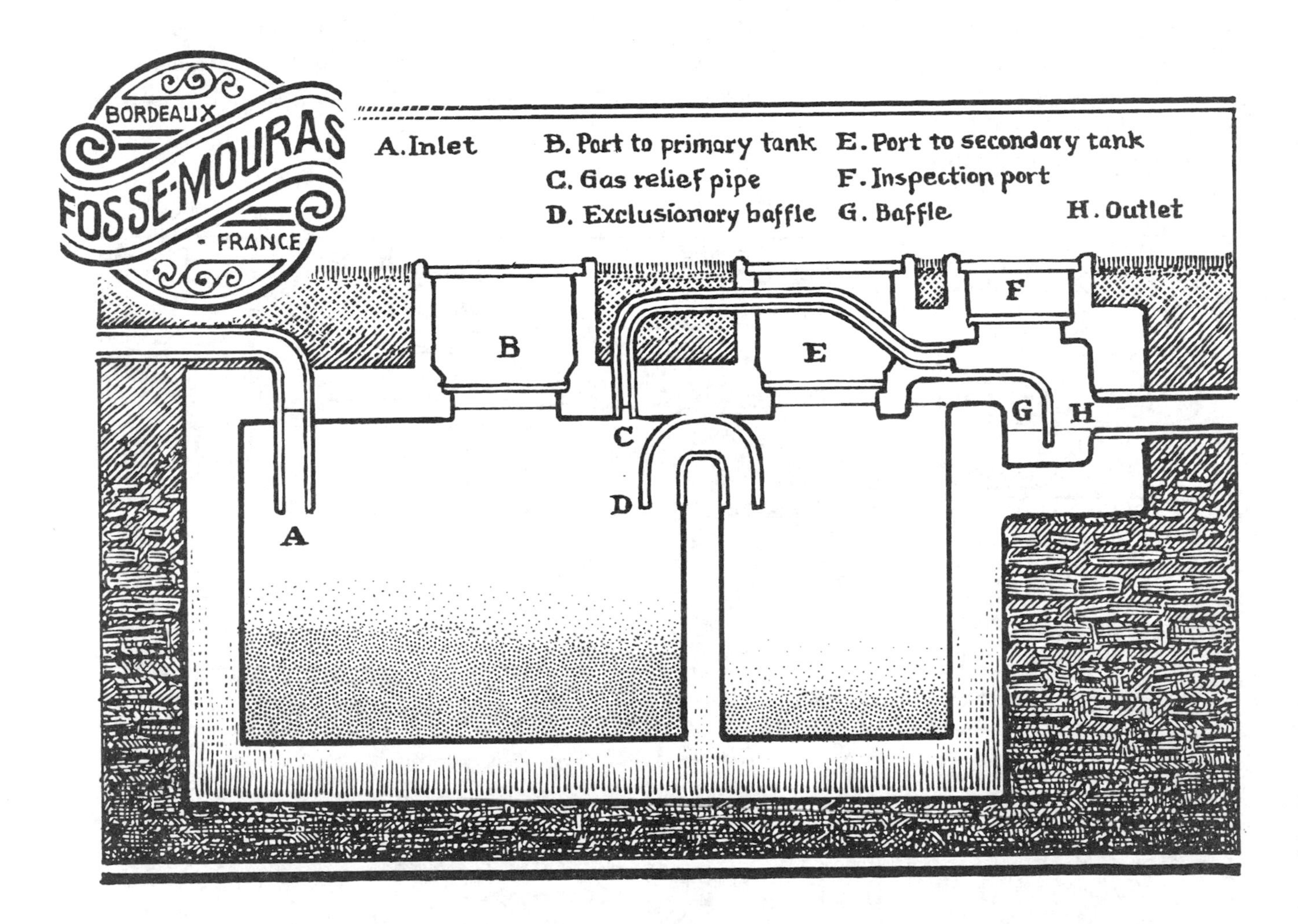

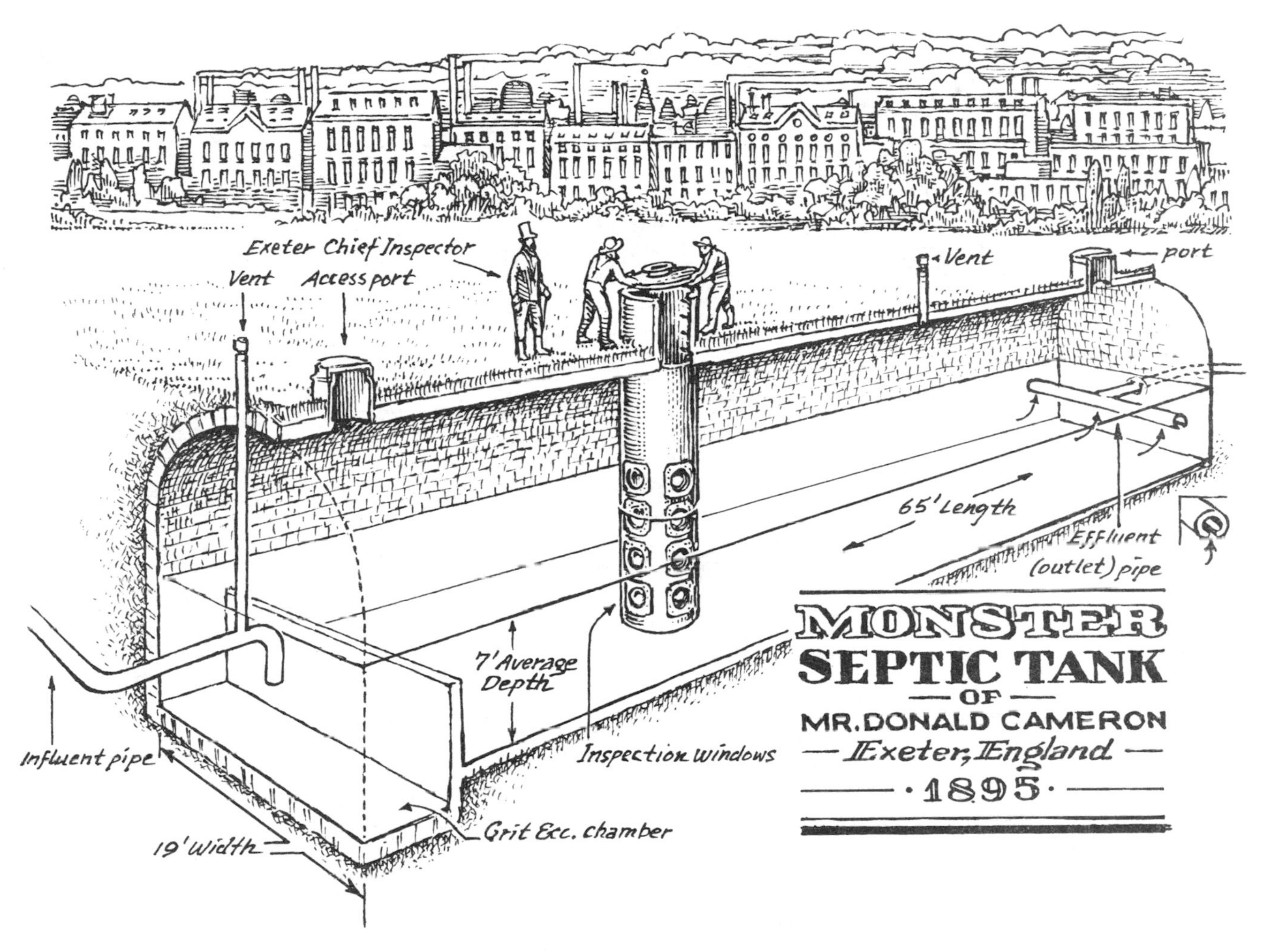

Donald Cameron's 53,000-gallon septic tank was hardly "monstrous" compared to the Fosse-Mouras tank, or the truly monstrous tank in Birmingham, England that covered 5 acres. Cameron is credited with coining the name "septic tank."

COPYRIGHT DISPUTES

Engineers in the early 1900s understood a fair amount about the functions of the septic tank. They recognized it as a valid, primary sewage-treatment device, even though secondary treatment (purification of effluent) was not being provided. (There were no drainfields then.) Unfortunately, disputes over patents and royalties hindered use of septic tanks for community sewage treatment.

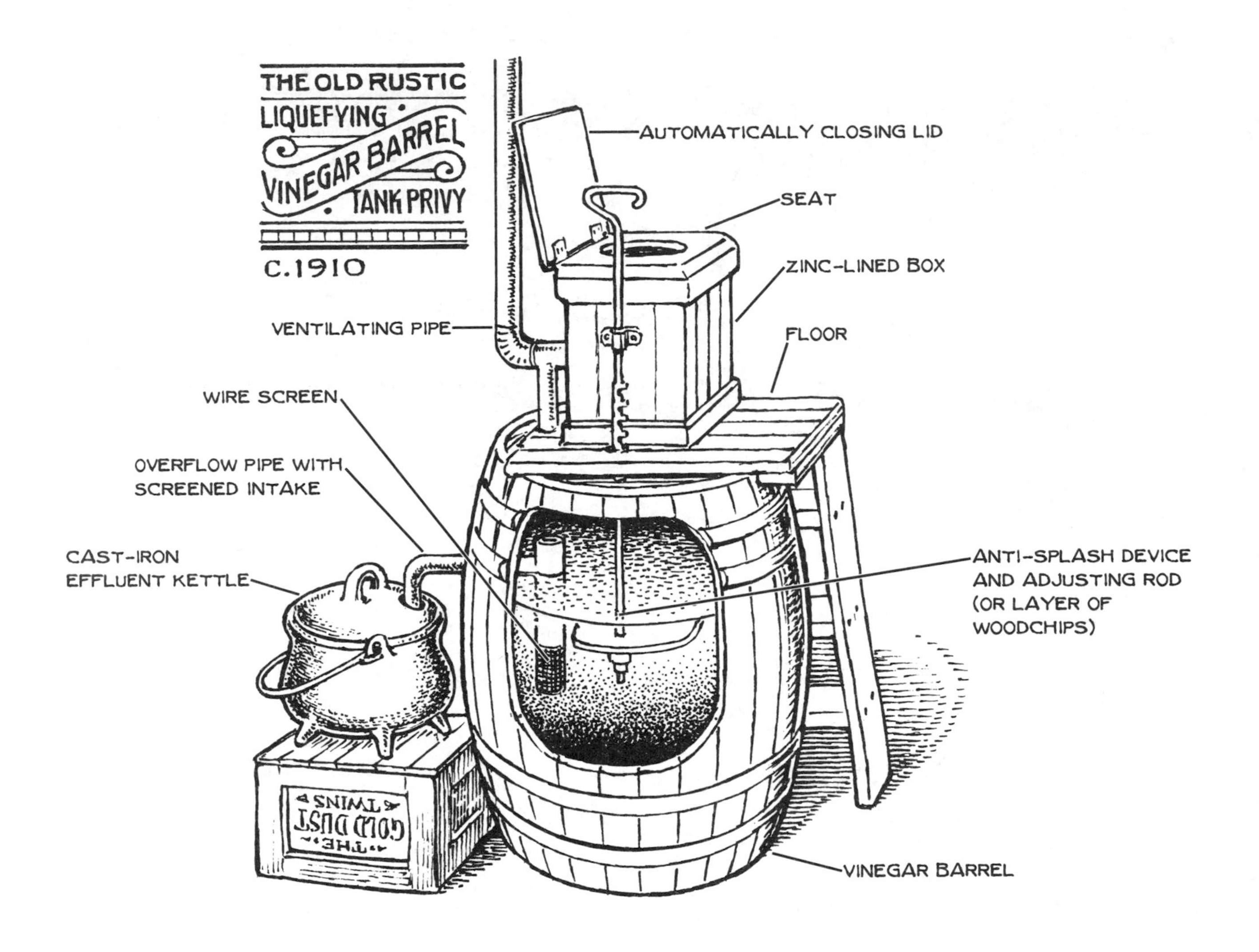

A U.S.-APPROVED PICKLE BARREL SEPTIC SYSTEM

In 1915, the U.S. Public Health Service (believe it or not!) designed a simple septic tank system consisting of a vinegar barrel as a tank and an iron pot as an effluent container.

As George Tchobanoglous points out in *Small and Decentralized Wastewater Management Systems,* "It should be noted that the use of a watertight tank and an effluent screen are two very important features of a modern septic tank."

EARLY DRAINFIELDS

In 1919, Leslie C. Frank and C. P. Rhymus of the U. S. Public Health Service did a feasibility study on the use of septic systems for single-family dwellings and small communities in the U.S. Frank and Rhymus recommended what was called the Imhoff tank to serve a single family household of five people. They also specified alternate drainfields fed by a distribution box—perhaps history's first dual drainfield! *(See pp. 20–21.)* Since this tank had no patent or royalty restrictions, it was adopted on a wide scale for single-family dwellings.

Over the years, tanks have changed in design somewhat, but they still perform the same functions with respect to reduction of solids and sludge. The major improvement in treatment of household wastes since the early 1900s has been the variety of soil absorption systems for treatment of septic tank effluent (sewage after it exits the tank). Most commonly this has taken the form of a drainfield.

Since the turn of the century, the use of septic tank systems for treatment of household wastewater has increased dramatically throughout the world.

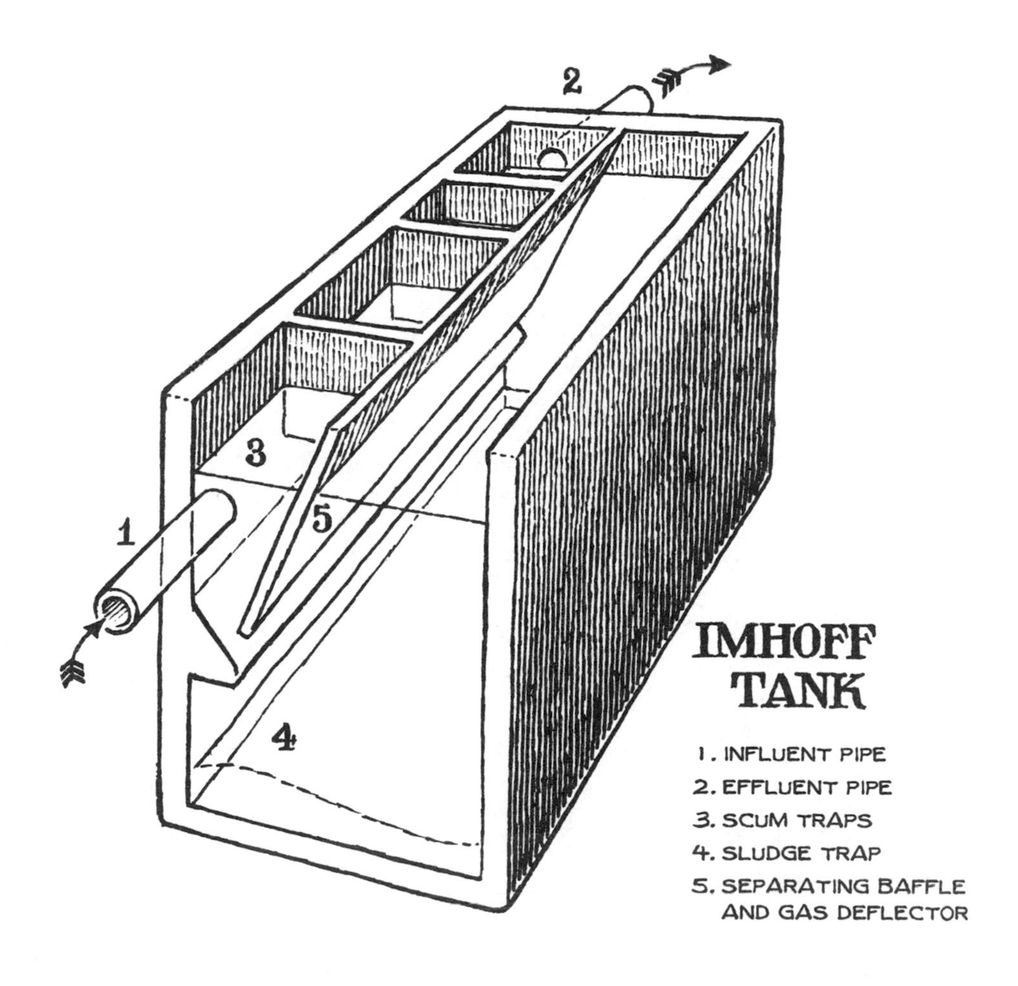

THE WALL STREET

Appendix

New Information

FOUR DIFFERENT TYPES OF SOIL

There has been a great deal of recent study on how the soil handles the waste and pathogens in the septic effluent. The EPA is now defining the drainfield or SWIS (Septic Wastewater Infiltration System) by the limiting conditions of the site in a much more thorough manner than simple percolation tests (how rapidly the soil absorbs the water). There must be adequate depth to what is called the "limiting condition." This refers to either an impermeable layer that blocks the flow of water to the soil beneath, or groundwater. (The shallower the limiting condition, the more difficult it is to achieve the proper degree of treatment.) With the appropriate system, as little as 12″–18″ of separation is sufficient. However, the distance is dependent not only on the system feeding the drainfield, but on the volume of liquid. For homeowners who want to be able to converse knowledgeably with health officials or engineers, the following chart from the EPA Onsite Wastewater Treatment Manual shows four different categories of permeable soil depth, and the system types suggested for each:

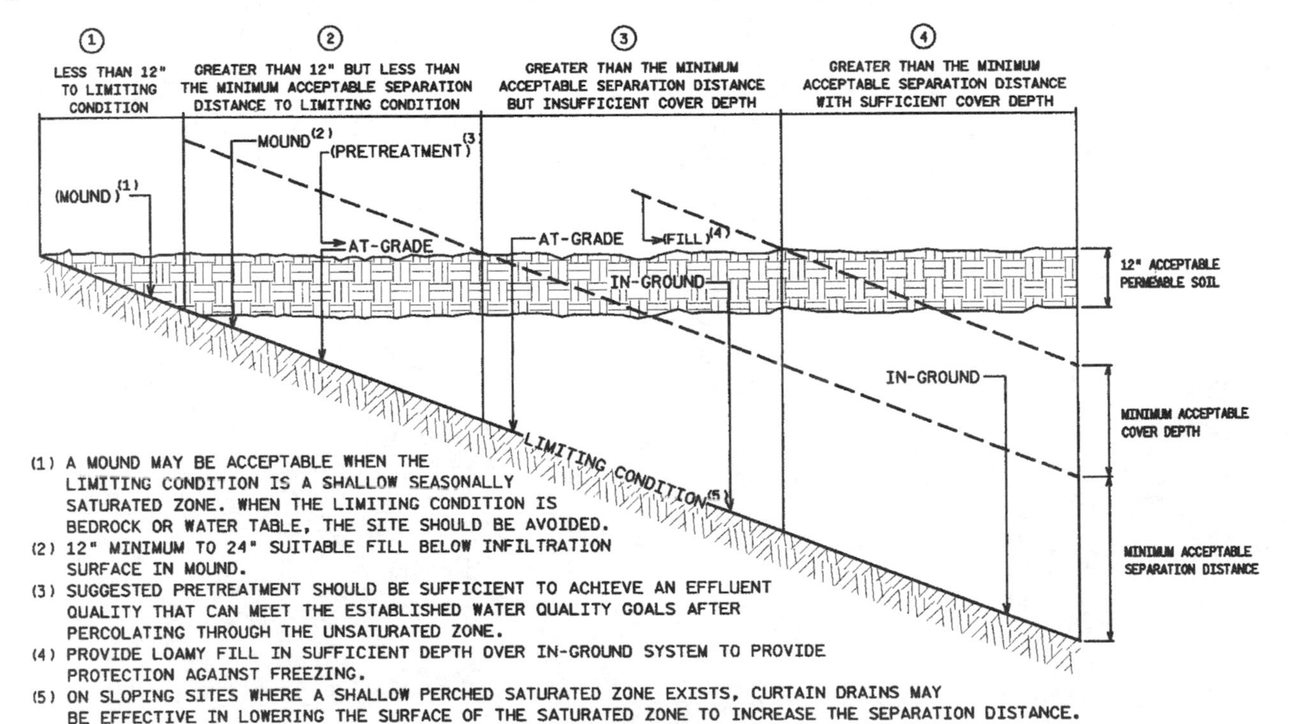

Type one soil, at left, is the worst condition shown here (the shallowest to the limiting condition). Likewise, type four is the best, i.e. the deepest to the limiting condition.

The entire manual is at:
http://www.epa.gov/ORD/NRMRL/Pubs/625R00008/625R00008.htm.

SOIL DRAINAGE

In addition to selecting the correct system for the soil depth of the site, it is also possible to improve the conditions with correct drainage (many old developments have not maintained their ditches and culverts, so soil tends to become overly saturated in wet weather, especially in flat areas). On sloped sites, curtain drains can be installed uphill of drainfields to keep out upslope water. The EPA manual references cases where level areas were drained of water with sump pumps. There is also a system using pipes to bleed air into the soil, maintaining an unsaturated area where soil bacteria can get enough oxygen to treat the waste.

Shallow Drainfields

Perhaps the most significant change since the first edition of this book is that most agencies, including the EPA, are giving the shallow drainfield, where all soil life resides, a lot more credit for the job they do in treating the waste. *(See pp. 22–23 on shallow drainfields.)* This is another reason the EPA places such emphasis on maintaining the vadose zone, where the soil receives oxygen from the air. This zone is essential to maintaining a healthy ecology of soil bacteria and maximizing uptake of nutrients and elimination of pathogens.

Pathway of Subsoil Aeration

Below is a cross section of a conventional (not shallow) gravel-filled drainfield trench and pipe system. This shows the pathways of air called the vadose zone, through the soil to the infiltration zone, where waste is treated.

From the EPA Onsite Wastewater Treatment Manual:
http://www.epa.gov/ORD/NRMRL/pubs/625r00008/625R00008chap4.pdf

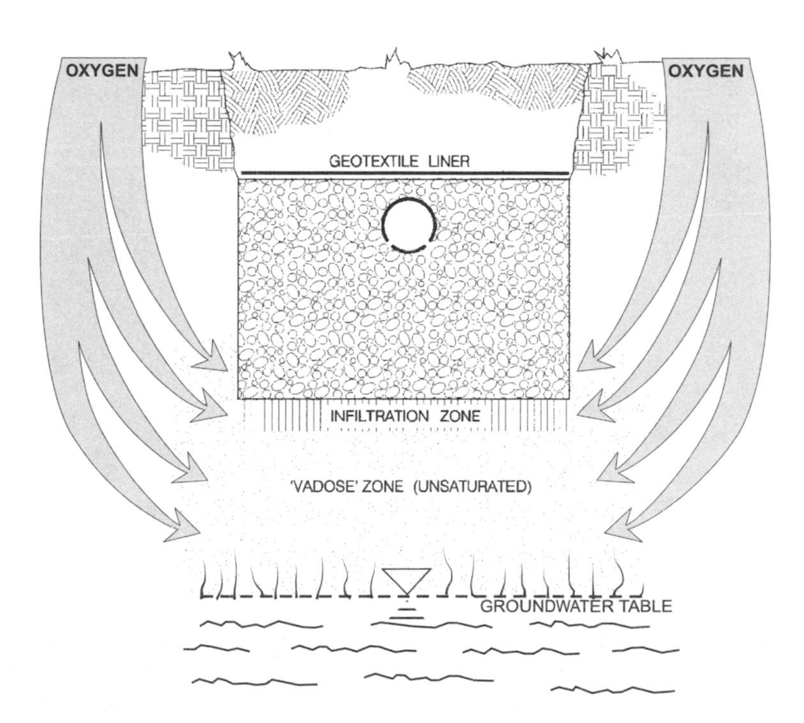

WHAT TREATMENT FOR WHAT OBJECTIVE?

Understanding the EPA chart on the opposite page means that when the health department or an engineer say you need to install a specific system, you can make sure that the treatment recommended is appropriate for the specific purpose. For example, if the objective is to remove nitrogen, you can see that an activated sludge system, such as Pirana™ *(see p. 97)*, or a recirculating media filter, such as Orenco's Advantex® system *(see p. 94)*, will accomplish that objective.

Commonly Used Treatment Processes and Optional Treatment Methods

It's helpful to know what the EPA recommends when you are dealing with health officials and engineers. The right-hand column of this chart shows many different approaches, depending on what the objective is.

From the EPA Onsite Wastewater Treatment Manual:

http://www.epa.gov/ORD/NRMRL/pubs/625r00008/625R00008chap4.pdf

<table>
<tr><th>TREATMENT OBJECTIVE</th><th>TREATMENT PROCESS</th><th>TREATMENT METHODS</th></tr>
<tr><td rowspan="2">SUSPENDED SOLIDS REMOVAL</td><td>Sedimentation</td><td>Septic tank
Free-water surface-constructed wetland
Vegetated submerged bed</td></tr>
<tr><td>Filtration</td><td>Septic tank effluent screens
Packed-bed media filters (incl. closed systems)
  Granular (sand, gravel, glass, bottom ash)
  Peat, textile
Mechanical disk filters
Soil infiltration</td></tr>
<tr><td rowspan="3">SOLUBLE CARBONACEOUS BOD AND AMMONIUM REMOVAL</td><td>Aerobic, suspended-growth reactors</td><td>Extended saturation
Fixed-film activated sludge
Sequencing batch reactions (SBRs)</td></tr>
<tr><td>Fixed-film aerobic bioreactor</td><td>Soil infiltration
Packed-bed media filters (incl. closed systems)</td></tr>
<tr><td>Lagoons</td><td>Facultative and aerobic lagoons
Free-water surface-constructed wetlands</td></tr>
<tr><td rowspan="2">NITROGEN TRANSFORMATION</td><td>Biological
  Nitrification
  Denitrification</td><td>Activated sludge (N)
  Sequencing batch reactors (N)
Fixed-film bio-reactor (N)
  Recirculating media filter (N, D)
  Fixed-film activated sludge (N)
  Anaerobic upflow filter (N)
  Anaerobic submerged media reactor (D)
Submerged vegetated bed (D)
Free-water surface-constructed wetlands (N, D)</td></tr>
<tr><td>Ion exchange</td><td>Cation exchange (ammonium removal)
Anion exchange (nitrate removal)</td></tr>
<tr><td rowspan="2">PHOSPHORUS</td><td>Physical/Chemical</td><td>Infiltration by soil or other media
Chemical flocculation and settling
Iron-rich packed-bed media filter</td></tr>
<tr><td>Biological</td><td>Sequencing batch reactions</td></tr>
<tr><td rowspan="2">PATHOGEN REMOVAL (bacteria, viruses, parasites)</td><td>Filtration/Predation/ Inactivation</td><td>Soil infiltration
Packed-bed media filters
  Granular (sand, gravel, glass, bottom ash)
  Peat, textile</td></tr>
<tr><td>Disinfection</td><td>Hypochlorite feed
Ultraviolet light</td></tr>
<tr><td rowspan="3">GREASE REMOVAL</td><td>Flotation</td><td>Grease trap
Septic tank</td></tr>
<tr><td>Adsorption</td><td>Mechanical skimmer</td></tr>
<tr><td>Aerobic biological treatment (incidental removal will occur; overloading is possible)</td><td>Aerobic biological systems</td></tr>
</table>

Glossary

Archimedes' principle: The principle of buoyancy, stating the resultant force on a wholly or partly submerged body acts vertically upward through the center of gravity of the displaced fluid and is equal to the weight of the fluid displaced.

absorption: The process by which one substance is physically taken into and included with another substance.

adsorption: The adhesion of molecules to the surface of solid bodies or liquids with which they are in contact.

aeration: The exposure to circulating air.

aerobic: Living, active, or occurring in the presence of oxygen.

aerobic bacteria: Bacteria that require free elemental oxygen to sustain life.

aerobic digestion: The breakdown of suspended and dissolved organic matter in the presence of oxygen. Usually associated with digestion of wastewater sludge.

aerobic lagoon: An oxygen-containing lagoon, often equipped with mechanical aerators, in which wastewater is partially stabilized by the metabolic activities of bacteria and algae. Small lagoons (less than ½ acre and less than 3 feet deep) may remain aerobic without mechanical aeration.

algae: Photosynthetic microscopic plants, which in excess can contribute taste and odor to potable water and deplete dissolved oxygen on decomposition.

alluvial: Relating to material deposited by flowing water.

amoeba: A usually microscopic one-celled animal or protozoan motile by means of pseudopodia.

anaerobic: Living, active, or occurring in the absence of oxygen.

anaerobic bacteria: Bacteria that grow only in the absence of free elemental oxygen.

anaerobic digestion: The degradation of organic matter brought about through the action of microorganisms in the absence of elemental oxygen.

anaerobic lagoon: A wastewater or sludge treatment process that involves retention under anaerobic conditions.

anaerobic waste treatment: Waste stabilization brought about through the action of microorganisms in the absence of air or elemental oxygen. Usually refers to waste treatment by methane fermentation.

aquifer: A porous, water-bearing geologic formation. Generally restricted to materials capable of yielding an appreciable supply of water.

backfill: (1) The operation of refilling an excavation, usually after some structure has been placed therein. (2) The material placed in an excavation in the process of backfilling.

backflow-prevention device: Any effective device, method, or construction used to prevent backflow into a potable water system.

backwash: The reversal of flow through a rapid sand filter to wash clogging material out of the filtering medium and reduce conditions causing loss of head. Applies to process of removing diatomaceous earth filter prior to pre-coating.

bacteria: A group of universally distributed, rigid, essentially unicellular, microscopic organisms lacking chlorophyll. They perform a variety of biological treatment processes, including biological oxidation, sludge digestion, nitrification, and denitrification.

bacterial analysis: The examination of water and wastewater to determine the presence, number, and identity of bacteria; more commonly called bacterial examination.

bacteriological count: A means for quantifying numbers of organisms.

baffle aerator: An aerator in which baffles are provided to cause turbulence and minimize short-circuiting.

baffles: Deflectors, vanes, guides, grids, gratings, or similar devices constructed or placed in flowing water, wastewater, or slurry systems as a check or to effect a more uniform distribution of velocities; absorb energy; divert, guide, or agitate the liquids; and check eddies.

ball joint: A flexible pipe joint made in the shape of a ball or sphere.

ball valve: A simple non-return valve consisting of a ball resting on a cylindrical seat within a fluid passageway.

bedrock: The solid rock encountered below the mantle of loose rock and more or less unconsolidated material which occurs on the surface of the earth's crust. In many places, bedrock appears at the surface.

biochemical: (1) Pertaining to chemical change resulting from biological action. (2) A chemical compound resulting from fermentation. (3) Pertaining to the chemistry of plant and animal life.

biochemical action: Chemical change resulting from the metabolism of living organisms.

biochemical process: The process by which the metabolic activities of bacteria and other microorganisms break down complex organic materials into simple, more stable substances.

biodegradation: The destruction of organic materials by microorganisms, soil, natural bodies of water, or wastewater treatment plants.

biological filter: A bed of sand, gravel, broken stone, or other medium through which wastewater flows or trickles, which depends on biological action for effectiveness.

biomat: The *biological mat* (biomat) is a black, jelly-like mat about one to two inches thick, that forms at the gravel-soil interface at the bottom and sidewalls of the drainfield trench. The biomat is composed of microorganisms (and their byproducts) that anchor themselves to soil and rock particles, and whose food is the organic matter in the septic tank effluent. Since the biomat has a low permeability, it serves as a valve to slow down and control the rate of flow out of the trench into the drainfield soil, and also serves as a filter to provide effluent treatment.

blackwater: Wastewater which comes from either the toilet or the kitchen sink. (Kitchen sink water is not universally considered to be blackwater, but because of the many raw food particles that it contains and the problems that they can cause, we classify it as blackwater.)

building drain: In plumbing, that part of the lowest horizontal piping within a building that conducts water, wastewater, or storm water to a building sewer.

building sewer: In plumbing, the extension from the building drain to the public sewer or other place of disposal. Also called *house connection.*

cesspool: A lined or partially lined underground pit into which raw household wastewater is discharged and from which the liquid seeps into the surrounding soil. Sometimes called *leaching cesspool.*

chlorination: The application of chlorine or chlorine compounds to water or wastewater, generally for the purpose of disinfection, but frequently for chemical oxidation and odor control.

clarification: Any process or combination of processes, the primary purpose of which is to reduce the concentration of suspended matter in a liquid. Term formerly used as synonym of *settling* or *sedimentation.* In recent years latter terms preferable when describing settling process.

clarified wastewater: Wastewater from which most of the settleable solids have been removed by sedimentation. Also called *settled wastewater.*

cleanout: Any structure or device which is designed to provide access for the purpose of removing deposited or accumulated materials.

coliform: One type of bacteria. The presence of coliform-group bacteria is an indication of possible pathogenic bacterial contamination. *Fecal coliforms* are those coliforms found in the feces of various warm-blooded animals, whereas the term *coliform* also includes other environmental sources.

coliform-group bacteria: A group of bacteria predominantly inhabiting the intestines of humans or animals, but also occasionally found elsewhere. It includes all aerobic and facultative anaerobic, gram-negative, non-spore–forming, rod-shaped bacteria that ferment lactose with production of gas. Also included are all bacteria that produce a dark, purplish-green metallic sheen by the membrane filter technique used for coliform identification. The two groups are not always identical, but they are generally of equal sanitary significance.

compost: The product of thermophilic biological oxidation of sludge or other materials.

denitrification: The anaerobic biological reduction of nitrate nitrogen to nitrogen gas. Also removal of total nitro gen from a system.

detention time: The period of time that a water or wastewater flow is retained in a basin, tank, or reservoir for storage or completion of physical, chemical, or biological reaction.

digested sludge: Sludge digested under either aerobic or anaerobic conditions until the volatile content has been reduced to the point at which the solids are relatively nonputrescible and inoffensive.

digestion: (1) The biological decomposition of organic matter in sludge, resulting in partial liquefaction, mineralization, and volume reduction. (2) The process carried out in a digester.

discharge: With respect to wastewater, the release of treated, partially treated, or untreated wastewater into the environment, whether accidentally or by design.

disposal: Release to the environment.

domestic wastewater: Wastewater derived principally from dwellings, business buildings, institutions, and the like. It may or may not contain groundwater, surface water, or storm water.

ecosystem: The composite balance of all living organisms and plants and the ambient environment; air, water, and solid phases; in a defined area.

effluent: Wastewater flowing out of a septic tank.

evapotranspiration: Water withdrawn from soil by evaporation and/or plant transpiration. Considered synonymous with consumptive use.

fecal coliform: Aerobic and facultative, gram-negative, non-spore–forming, rod-shaped bacteria capable of growth at 44.5° C, and associated with fecal matter of warm-blooded animals.

feces: Excrement from the gastrointestinal tract, consisting of residue from food digestion and bacterial action.

filter: A device or structure for removing solid or colloidal material, usually of a type that cannot be removed by sedimentation, from water, wastewater, or other liquid. The liquid is passed through a filtering medium, usually a granular material but sometimes finely woven cloth, unglazed porcelain, or specially prepared paper. There are many types of filters used in water and wastewater treatment.

filtration: The process of contacting a dilute liquid suspension with filter media for the removal of suspended or colloidal matter or for the dewatering of concentrated sludge.

flocculate: To cause (soil) to form small lumps or masses.

French drain: An underground passageway for water through the interstices among stones placed loosely in a trench.

gravity system: (1) A system of conduits (open or closed) in which the liquid runs on descending gradients from source to outlet and where no pumping is required. (2) A water distribution system in which no pumping is required.

graywater: Wastewater which does not come from the toilet or kitchen sink.

graywater system: A system which separates graywater from blackwater to divert it away from the septic tank.

grease interceptor (trap): In plumbing, a receptacle designed to collect and retain grease and fatty substances normally found in kitchen or similar wastes. It is installed in the drainage system between the kitchen or other point of production of the waste and the building sewer.

groundwater: Subsurface water occupying the saturation zone from which wells and springs are fed. In a strict sense the term applies only to water below the water table.

groundwater recharge: Replenishment of groundwater naturally by precipitation, runoff, or artificially by spreading or injection.

household wastes: The water-carried wastes from kitchens, toilets, lavatories, and laundries.

infectious hepatitis: An acute viral inflammation of the liver characterized by jaundice, fever, nausea, vomiting, and abdominal discomfort. May be waterborne.

infiltration: (1) The flow or movement of water through the interstices or pores of soil or other porous medium. (2) The quantity of groundwater that leaks into pipe through joints, porous walls, or breaks. (3) The entrance of water from the ground into a gallery. (4) The absorption of liquid by the soil either as it falls as precipitation or from a stream flowing over the surface.

leachate: Liquid that has percolated through solid waste or other permeable material and has extracted soluble, dissolved, or suspended materials from it.

leaching: (1) The undesirable removal of soluble constituents from soil, landfills, mine wastes, sludge deposits, or other material by percolating water. (2) The desirable disposal of excess liquid through porous soil or rock strata.

mastic: Any of various pasty materials used as protective coatings or cements.

microorganism (microbe)**:** A living creature of microscopic or submicroscopic size.

nitrification: The oxidation of ammonia nitrogen to nitrate nitrogen in wastewater by biological or chemical reactions.

nitrogen: An essential nutrient that is often present in wastewater as ammonia, nitrate, nitrite, or organic nitrogen. The concentrations of each form and the sum total nitrogen are expressed as mg/l elemental nitrogen. Also present in some groundwater as nitrate and in some polluted groundwater in other forms.

nutrient: A substance that nourishes living things: for example, nitrogen and phosphorus in sewage nourish green plants.

organic matter: A material made up of or derived from living things.

package treatment plant: A small wastewater treatment plant, often fabricated at the manufacturer's factory, hauled to the site, and installed as one facility. The package may be either a small primary or a secondary wastewater treatment plant.

pathogens: Organisms (mostly microbes) that cause diseases.

percolation: Seepage through a permeable material.

permeable: Having pores or openings that permit liquids or gases to pass through.

primary effluent: The liquid portion of wastewater leaving primary treatment.

primary sludge: Sludge obtained from a primary settling tank.

primary treatment: (1) The first major treatment in a wastewater treatment facility, usually sedimentation but not biological oxidation. (2) The removal of a substantial amount of suspended matter but little or no colloidal and dissolved matter. (3) A wastewater treatment process usually consisting of clarification with or without chemical treatment to accomplish solid-liquid separation.

protozoa: Small one-celled animals including amoebae, ciliates, and flagellants.

putrefaction: Biological decomposition, usually of organic matter, with the production of ill-smelling products associated with anaerobic conditions.

Salmonella: A genus of aerobic, rod-shaped, usually motile bacteria that are pathogenic for man and other warm-blooded animals.

sand filter: A bed of sand through which water is passed to remove fine suspended particles. Very common in water treatment plants; also used in tertiary wastewater treatment plants and sludge drying beds.

scum (cake)**:** A layer of wastewater particles floating on the liquid surface in a septic tank.

septage: The sludge produced in individual onsite wastewater disposal systems such as septic tanks and cesspools.

septic tank: An underground vessel for treating wastewater from a single dwelling or building by a combination of settling and anaerobic digestion. Effluent is usually disposed of by leaching. Settled solids are pumped out periodically and hauled to a treatment facility for disposal.

septic wastewater: Wastewater undergoing anaerobic decomposition.

settleable (floatable) solids: (1) That matter in wastewater which does not stay in suspension during a preselected settling period, such as one hour, but settles to the bottom. (2) Suspended solids that can be removed by conventional sedimentation.

settling: The process of subsidence and deposition of suspended matter carried by water, wastewater, or other liquids, or by gravity. It is usually accomplished by reducing the velocity of the liquid below the point at which it can transport the suspended material; also called *sedimentation*.

sewage: Household and commercial wastewater that contains human waste. Distinguished from industrial wastewater.

sewer: A pipe or conduit that carries wastewater or drainage water.

sewer system: Collectively, all of the property involved in the operation of a sewer utility. It includes land, wastewater lines and appurtenances, pumping station, treatment works, and general property. Occasionally referred to as a *sewerage system*.

sludge: The accumulated solids that have settled to the bottom of a septic tank.

sludge digestion: The process by which organic or volatile matter in sludge is gasified, liquefied, mineralized, or converted into more stable organic matter through the activities of either anaerobic or aerobic organisms.

soil absorption capacity: In subsurface effluent disposal, the ability of the soil to absorb water.

solids: In water and wastewater treatment, any dissolved, suspended, or volatile substance contained or removed from water or wastewater.

STEP system: Septic tank effluent pumping system, where each house in a community has a tank, with effluent pumped to a central drainfield.

subsurface wastewater disposal: The process of wastewater treatment and disposal in which wastewater or effluent is applied to land by distribution beneath the surface through open-jointed pipes or drains.

surface aeration: The absorption of air through the surface of a liquid.

suspended solid: Solid material suspended in wastewater.

SWIS: Subsurface wastewater infiltration system; the new name for drainfields or leachfields.

tank: A structure or container for containing water or wastewater for purposes such as aeration, disinfection, equalization, holding, sedimentation, or treatment; or for mixing, dilution, feeding, or other handling of chemical additives.

tank treatment: The detention of wastewater in tanks, either quiescent or with continuous flow.

tertiary treatment: The treatment of wastewater beyond the secondary or biological stage. Term normally implies the removal of nutrients, such as phosphorus and nitrogen, and a high percentage of suspended solids. Term now being replaced by the preferred term, *advanced waste treatment*.

toxicity: The adverse effect which a biologically active substance has, at some concentration, on a living entity.

trace element: Any element in water or wastewater that, for reasons associated with natural distribution, industrial use, solubility, or other factors, is present at very low concentrations.

trap: (1) A device used to prevent a material flowing or carried through a conduit from reversing its direction of flow or movement or from passing a given point. (2) A device to prevent the escape of air from sewers through a plumbing fixture or catch basin.

treated sewage: Wastewater that has received partial or complete treatment.

underdrain: A drain that carries away groundwater or the drainage from prepared beds to which water or wastewater has been applied.

virus: The smallest (10 to 300 mμ in diameter) life form capable of producing infection and diseases in man or in other large species.

wastewater: The spent or used water of a community or industry which contains dissolved and suspended matter.

wastewater disposal: The act of disposing of wastewater by any method. Not synonymous with *wastewater treatment*. Common methods and instruments of disposal are: dispersion, dilution, broad irrigation, privy cesspool.

wastewater reclamation: Processing of wastewater for reuse.

waterborne disease: A disease caused by organisms or toxic substances carried by water. The most common waterborne diseases are typhoid fever, Asiatic cholera, hepatitis, giardiasis, dysentery, tetanus, polio, and other intestinal disturbances.

zero discharge: (1) Complete recycling of water. (2) Discharge of essentially pure water. (3) Discharge of a treated effluent containing no substance at a concentration higher than that found normally in the local environment.

Bibliography

SERIOUS BOOKS ON SEPTIC SYSTEMS

Small and Decentralized Wastewater Management Systems
Ron Crites and George Tchobanoglous
McGraw Hill, Inc., New York, NY, 1998

The bible of small-scale wastewater engineering, this is a 1000-page textbook for both students and engineers, and contains the most up-to-date information available on decentralized wastewater treatment systems. It is more than a homeowner needs, but is a highly valuable source of engineering and scientific details for professionals and practitioners, as well as for small towns and wastewater districts. Since "more than 60 million people in the United States live in homes that are served by decentralized collection and treatment systems . . . decentralized wastewater management becomes of great importance for future management of the environment." Although it is full of charts, formulas, and complex engineering data, the text is clear and informative, and the sections on septic tanks, alternative wastewater collection systems, wetlands, and recirculating sand and gravel filters will be of interest to people with failing systems and/or to environmentally conscious wastewater districts.

Environmental Engineering and Sanitation, Fourth Edition
Joseph A. Salvato
John Wiley & Sons, Inc., New York, NY, 1992

This book, first published in 1958 (and since updated), is "devoted to the study of the quality of the environment and to the technology of its conservation" and is a 1300-page treatise for teachers, students, and professionals in public health and environmental protection. It includes a comprehensive treatment of 12 topics of environmental concern, most of which are not related to wastewater management. However, the section on small wastewater disposal systems is very clearly written and contains a number of excellent drawings of drainfields, drainage systems, and alternative systems such as mounds and sand filters. The book stresses practical solutions to environmental problems and is a valuable resource.

Septic-Tank Systems, A Consultant's Toolkit
John H. Timothy Winneberger
Butterworth Publishers, Stoneham, MA, 1984

The author is a former plant physiologist who, in 1958, researched septic systems for the FHA. The study of septic systems eventually became "my life's work and improving practices became a very personal, almost religious cause." Winneberger, now retired and living in New Mexico, has played a very important role in the understanding of wastewater science in the U.S. He has constantly advocated the desirability of small-scale septic systems as opposed to centralized sewers, and champions the functionality of well-designed, onsite systems. In this book he says he prefers to "emulate the relaxed style of early scientists," leaving the discriminating reader to "distinguish factual data from a writer's personal view." The book is divided into two parts: Part I deals with subsurface disposal of septic tank effluent; Part II addresses the septic tank. Part I deals extensively with the percolation test, which Winneberger claims is often incorrectly administered, or the results incorrectly interpreted. Both parts of the book are highly informative, but, in spite of the informal style, still highly technical. Winneberger is a practical philosopher and takes aim at what he calls "superstitions" about today's wastewater practices.

Onsite Wastewater Disposal
Richard J. Perkins
Lewis Publishers, Inc., Chelsea, MI, 1990

The purpose of this book is "to provide information sufficient to allow a homeowner, a potential homeowner, contractor, septic system installer, or consulting engineer to evaluate the future site of a liquid waste disposal system, to identify any potential problems, and to select and design a system that will provide adequate protection at minimal cost." Perkins provides an easy-to-read, informed discussion with a minimum of formulas. In addition to information on proper siting and soil characteristics, system design construction and maintenance, the author includes discussions of system modifications (mounds, sand filters, wetlands) and composting toilets and graywater. Recommended reading for those with any interest in onsite systems.

Onsite Wastewater Treatment Systems
Bennette Day Burks and Mary Margaret Minnis
Hogarth House, Ltd., Madison, WI, 1994

The authors, both with scientific backgrounds and trained in wastewater management, purchased homes served by traditional onsite systems—only to be surprised by the dearth of information on the topic. They determined to write a book that would bridge the information gaps between the various groups involved in wastewater practices, from municipal wastewater treatment systems to manufacturers, regulators, and academics. The book is written for science or engineering students, regulators, policymakers, and interested citizens, and given the intended audience, it is no surprise that it is fairly technical, with illustrative charts and formulas. However, the average homeowner will find the book readable and informative, with interesting sections on the properties of water, on microbiology, and on dealing with the variety of wastewater regulations and regulators nationwide.

Septic Systems Handbook
O. Benjamin Kaplan
Lewis Publishers, Inc., Chelsea, MI, 1987

A clear, concise, practical, slightly quirky handbook from a soils scientist and registered sanitarian in San Bernardino County, California. The author is concerned with diseases caused by improper disposal of sewage and stresses the importance of public health agencies in controlling disposal of wastewater or sewage. There is a brief description of septic tanks and drainfields, but the heart of the book is about soils, proper percolation tests, and consequences of improper disposal of sewage. Intended for professionals who deal with septic systems, but clear enough for the serious homeowner.

HOME & HOMESTEAD BOOKS

Septic Tank Practices
Peter Warshall
Anchor Press/Doubleday, New York, 1979

Septic Tank Practices (unfortunately out of print) was the first book about onsite systems written entirely for the layperson. It resulted from the residents of a small coastal town uniting to defeat a costly and over-planned centralized sewerage system that would have created a potentially massive build-out of a bucolic rural area. It is an informative as well as a charming book, which stresses the effectiveness of natural and biological wastewater treatment. It was years ahead of its time in dealing with the ecological impact of our culture's treatment of sewage—something that has now made its way into mainstream textbooks. It has a unique chapter on soil called "Good Soil, Clean Water," which describes the wondrous powers of microorganisms in soil, and how soil acts to filter and clean pathogens and viruses. There is a thread of do-it-yourself design advice running throughout the book, from configuration of septic tanks, to profiles of drainfields and mounds, to doing your own soil and percolation tests.

Wells and Septic Systems
Max and Charlotte Alth,
Revised by S. Blackwell Duncan
Tab Books/McGraw Hill, Inc., New York, NY, 1992

This is a book for people who must (or want to) dig their own well and/or build their own septic system. The first third of the book deals with septic systems, and begins with a short chapter on how septic systems work; its description of the microscopic action in the tank and soil is excellent, succinct, and clear. After pointing out that there are five million bacteria in a teaspoon of ordinary soil, the authors describe the dance of microbes, nematodes, and the like in killing off viruses in wastewater. There are then detailed descriptions for building a tank out of concrete block or casting it in place, working with sewer pipe, and constructing various types of drainfields. (*Note:* Building a septic tank is not a project for someone with no construction experience.) The last part of the book is a detailed description of how to dig, drill, jet, or bore a well.

Cottage Water Systems:
An Out-Of-The-City Guide to Pumps, Plumbing, Water Purification, and Privies
Max Burns
Cottage Life Books, Toronto, Ontario, Canada, 1993

This book was written for folks who have a vacation home or a second home with a private water system. It covers how the different components of a home water system work, and there are tips on installation and repair as well as troubleshooting advice if things go wrong. The illustrations in this book are excellent. Topics covered include sources of water, types of pumps and where to get them, hooking up a system, water quality, water purification, septic systems, outhouses, alternative toilets, graywater, closing and opening (the cottage), and getting water in winter.

Chelsea Green Publishing Company
White River Junction, VT 800-296-6300
http://www.chelseagreen.com/

Chelsea Green books will be of interest to those living in the country and/or interested in providing at least a part of their own food, energy, or shelter. Chelsea Green has produced a line of well-designed and useful books on homes and the home arts, including *The Straw Bale House* by Athena and Bill Steen and David Bainbridge, and *The Shelter Sketchbook* by John Taylor. On their website is the statement: "Chelsea Green publishes information to help us design future lives where human activities of production and consumption are balanced in harmony with the natural state of the world." Amen.

WATERLESS WASTE DISPOSAL

The Composting Toilet System Book
David Del Porto and Carol Steinfeld
The Ecowaters Project
Concord, MA, 1999
http://www.ecowaters.org

A landmark of a book, long overdue. This is a very thorough, complete, carefully researched book on composting toilet systems. Over 50 systems are described, including manufactured units such as the SunMar, Biolet, Phoenix, Clivus, and Carousel. There are also numerous ideas and construction details for owner-built systems. How to install the various systems, and most importantly, how to maintain them. The advantages and disadvantages of each system are listed. Many composting toilets promoted and built (both manufactured and home-built) 20 years ago had serious flaws, so the concept has a dubious reputation with people who encountered smelly and malfunctioning units in those days. Designers and manufacturers learned from previous mistakes, however, and the technology has greatly improved. Read all about it here; it's essential if you're interested in the subject.

Excreta Disposal for Rural Areas and Small Communities
E. G. Wagner and J. N. Lanoix
World Health Organization, Geneva, 1958

Approximately 70 percent of this book, which was written by WHO sanitary health engineers, concerns privies and is of interest to anyone considering constructing a privy. The authors bring a world view to this book and, therefore, discuss options that would hardly be approved by health officials in the '90s in the United States, such as the overhung latrine—(overhanging the stream or lake which conducts the raw effluent away)—and note that this solution, as undesirable as it is, is sometimes the only option in areas of Asia where many people are forced to inhabit land areas that are periodically covered with water. The authors make clear the connection between improper handling of excreta and disease. The fact that the book is 40 years old explains why the authors are less concerned than public health officials today about the possible contamination of groundwater or large bodies of water by excreta disposal, but other than this, the book is still timely in its discussion of various solutions for rural sanitation. After reading the various pros and cons of a variety of privies—pit privy, aqua privy (sludge has to be "bailed out" periodically), water seal latrine, bored-hole latrine, bucket latrine, French latrine (dig a hole, fill it up, cover it, and move on), one reaches the section on composting toilets, septic tanks, and drainfields with a sigh of relief. Composting toilets have come a long way since this book was written. All of these systems are predicated on getting "the family to assume responsibility for, and to solve, its own excreta disposal problems" —and there's the rub!

SAVING WATER

The New Create an Oasis with Greywater: Revised and Expanded 5th Edition
Art Ludwig
Oasis Design Press, Santa Barbara, CA, 1995
http://www.greywater.net/

Art Ludwig has been focusing on graywater disposal for a decade or more now, and in this 5th edition of his manual, he has incorporated many new features, including branched drains

—the network of pipes used to disperse graywater via gravity to different locations in the soil. There are new drawings, new systems, descriptions of "lessons learned" (what turned out not to work), details on site assessment, including how to shoot levels either with a home-made bucket / hose level, or a transit. The book is full of friendly drawings and clear photos describing many aspects of functional graywater disposal.

Gray Water Use in the Landscape:
How to Use Gray Water to Save Your Landscape During Droughts
Robert Kourik
Metamorphic Press, Santa Rosa, CA, 1988

This is a 28-page pamphlet (in its 10th printing) written and published by Robert Kourik, author of *Drip Irrigation for Every Landscape and All Climates. (See next review.)* There are 11 very clear line drawings showing you how to redirect water from your existing plumbing to a graywater system, details on how to plumb a barrel for a washing machine graywater system, and how to make a mini-drainfield. In the appendix is a list of safe soaps, detergents, and cleaners, and details on tools and techniques for working with plastic pipes. Kourik is a thoughtful steward of the earth's resources and an excellent writer, with the ability to make technical information understandable to the layperson. A list of his books is on the web at:

http://www.terrainforma.com/books.html

Drip Irrigation for Every Landscape and All Climates
Robert Kourik
Metamorphic Press, Santa Rosa, CA, 1992

This book is thoughtfully laid out, carefully explained, wonderfully illustrated, and has just the right sprinkling of humor. Robert Kourik has written a clear, concise, easy-to-read book that tells you everything you need to know to install and maintain a drip system. "I have distilled into this book what I've learned from thousands of hours of self-inflicted mistakes." Kourik says that instead of explaining a multitude of different ways of doing things, he'll explain the basic principles of design, familiarize you with the best parts, and pass on some tricks of the trade. He does just that in a clear format that begins with advantages and disadvantages, then moves to why drip irrigation works, with a discussion of soil and roots. He then describes basic drip irrigation, a basic project, when and how long to irrigate, and hiding and expanding a drip system. The book also covers special areas such as drip irrigation for containers, trees and shrubs, and vegetable beds.

VINTAGE STUFF

Flushed with Pride,
The Story of Thomas Crapper
Wallace Reyburn
Pavilion Books, Ltd., London, 1989

His name may be slang in America for a toilet (the term apparently brought back by American servicemen in Britain who saw it printed on the tanks of toilets), but in Victorian England, Thomas Crapper was a plumbing superstar. In a story that is reminiscent of a Dickens novel, Thomas was apprenticed at age 11 to a plumber in London and then walked 165 miles from Yorkshire to London to begin a career that was to bring him to the pinnacle of his profession: *Royal Plumber*. Along the way, this enterprising and inventive man not only perfected the flush toilet, but also secured numerous other patents such as an automatically flushing W.C., stair treads, self-rising closet seats (this one never achieved popularity—too expensive), and the "wall-hung" cantilevered toilet with a cistern and all pipes out of sight (first developed for prisons and mental institutions). Crapper's biography provides an amusing and often gossipy look at Victorian times, Victorian royalty, *and* Victorian plumbing. As his biographer remarks, "in his time if he didn't walk with Kings, he at least discussed sanitary arrangements with them."

Thunder, Flush and Thomas Crapper—an Encycloopedia
Adam Hart-Davis
Trafalgar Square Publishing, North Pomfret, VT, 1997

An encycloopedia *(sic)* of toilets, with alphabetical listing of all things lavatorial. Lots of great old drawings, interesting ephemera, and humorous historical anecdotes. A sketch of a pig loo in China, where the pigs, who hungrily wait underneath the privy chute for the feces, sometimes "jump up and snap hopefully at your bottom before you have finished."

Temples of Convenience and Chambers of Delight
Lucinda Lambton
St. Martin's Press, New York, NY, 1995

"Over 150 jewels of sanitation." A coffee table book with elegant color photos of English toilets, chamber pots, and bathroom accessories. Lucinda Lambton is an architectural photographer whose photographs are beautifully lit and carefully composed, and her introduction to the subject is witty and informative.

The Vanishing American Outhouse—A History of Country Plumbing
Ronald S. Barlow
Windmill Publishing Co., El Cajon, CA, 1989

There are a surprising number of books out there on toilets, privies, outhouses, and the like. Most of them are either weird or insubstantial. This one, however, is a solid, funny, and well-illustrated volume on the subject of American outhouses. There are over 100 photos, old steel engravings, humorous postcards, and construction diagrams that can be used to build different types of outhouses. By far the best book on the subject.

CURRENT INFO

Small Flows
The National Small Flows Clearinghouse
West Virginia University, P.O. Box 6064, Morgantown, WV 26506. 800-624-8301.
http://www.estd.wvu.edu.nsfc

Small Flows is a unique (free) quarterly publication on small community wastewater systems. The National Small Flows Clearinghouse is an invaluable source for communities seeking up-to-date wastewater information. They offer technical assistance—engineers who answer wastewater-related questions, databases and searches by request, and more than 250 informational and educational products, including books, brochures, case studies, and videotapes, which focus on small community wastewater treatment issues. This is the only publication of its kind anywhere.

SEPTIC SYSTEM INFORMATION ON THE INTERNET

The Consortium of Institutes for Decentralized Wastewater Treatment
http://centreforwaterresourcesstudies.dal.ca/pg2/page2.html

This is a group dedicated to providing information and encouraging research on decentralized, as opposed to centralized (i.e., sewers), wastewater systems. Lots of academic, government, and private sector links.

Infiltrator Systems Inc.
http://www.infiltratorsystems.com

Infiltrator makes drainfield modules *(see p. 23)* that can be used in drainfields, eliminating the need for gravel. Their website has forums for homeowners, engineers, and installers, as well as a wealth of septic system information, and links to other related sites.

Environmental Services and Training Division/ *Small Flows*
http://www.estd.wvu.edu/nsfc

West Virginia University's environmental website; it includes information on clean water as well as wastewater for small communities. It is also the gateway to WVU's *Small Flows* quarterly magazine online, with all kinds of info, research data, and relevant links. An excellent site.

Oasis Design
http://www.greywater.net

Art Ludwig is the inventor of Oasis soap, formulated to provide nutrient-laden graywater to garden plants from the washing machine. He is also the author of *Create an Oasis with Greywater. (See p. 169.)* This is his website, with information on graywater, water conservation, and biocompatible cleaning products. His publications are all listed, and useful data is shown from each one.

Orenco Systems, Inc.
http://www.orenco.com/

Orenco is an engineering and manufacturing firm dedicated to the development and production of high-quality wastewater treatment products for onsite treatment systems and pressure sewers. Their areas of expertise include intermittent and recirculating sand filters, shallow gravel-less drainfields, and effluent sewers. They manufacture a variety of products (including an effluent filter) for which they are highly regarded in the wastewater field.

Pumper Online
http://www.pumper.com

Pumper is a magazine for the liquid waste handling professional, and this is its website. It is a smart, witty source of info on equipment, supplies, parts, service, and education. One of its features is the *Septic System Answer Man.*

Septic Protector Home Page
http://www.septicprotector.com

These are the folks that produce the Septic Protector washing machine filter. *(See p. 175.)* They also have an excellent website with info on how septic systems work, causes of failure, and how to inspect the system of a house you are buying. They list a number of products other than their own, and have reader input and feedback. A valuable, intelligent, and witty source of information.

The Septic Systems Information Website
http://www.inspect-ny.com/septbook.htm

This is the best, most comprehensive website for septic system information. It has tons of info, references, FAQs, and a feature called "Ask the Master Plumber." There are many links to other sources.

Shelter Online
http://www.shelterpub.com
http://www.shelterpub.com/_shelter/ongoing_info.html

This is Shelter's website for the latest in septic system information. Since this book *(Septic Systems Owner's Manual)* presents only part of the picture, we have set up a section on our website to present updates, corrections, insights and feedback. It's a place for you to check out the latest, hear from other homeowners, and share any information you have developed or learned. Check us out on the web or email information or feedback related to septic systems to septic@shelterpub.com.

Terralift Soil Loosener
http://www.terraliftinternational.com

Failed drainfield? The Terralift is a machine that uses a probe and a pneumatic hammer to force air into and fracture the soil, restoring aerobic conditions. It injects polystyrene pellets into the soil to maintain passage for percolation. Costs range from $1000–$2000, but it can be cheaper than a new drainfield or entire system.

Zabel Environmental Design
http://www.zabel.com/

Zabel manufactures septic tank filters and has a playful website. They also produce *Zabel Zone*, a wastewater magazine that is viewable online via Adobe Acrobat. They state that their filters "remove solids from the discharged effluent, and promote regular maintenance by requiring periodic service." This means that as the filter gets clogged with grease, lint, etc. over the years, your drainage will start slowing down, alerting you to the fact that the filter needs cleaning. They also have a good question and answer section.

Grab Bag

Here are a few devices (and two suppliers) that we think you may find useful. We have not made an exhaustive survey of septic-related products, but we do know that the ones listed here are of high quality.

OUTLET TEE FILTERS

Orenco Systems, Inc., Sutherlin, OR
541-459-4449
http://www.orenco.com

Zabel Environmental Products, Louisville, KY
800-221-5742
http://www.zabel.com

The most serious problem with septic systems is a failed drainfield: when solids migrate from the tank into the drainfield, with resultant clogging. Two companies, Zabel and Orenco, make plastic filters that intercept solids at the outlet tee where the effluent flows into the drainfield. The filters need to be cleaned every two to three years, but are said to be somewhat self-cleaning if used properly. (The screens provide a surface area for microorganisms to grow, which in turn degrade the harmful particles into basic elements and compounds.) Both types of filters can be retrofitted into an existing septic tank and are especially valuable anywhere (residential or commercial) that wastewater has a high suspended solids content.

DRAIN KING

G. T. Water Products, Moorpark, CA
805-529-2900
http://www.gtwaterproducts.com/drain.html

This is a brilliant invention for unclogging plugged drains without using caustic chemicals such as Drano or calling a plumber. It comes in various sizes for different-size pipes and consists of a rubber bellows-type device that you connect to a garden hose. You insert it into the drain, turn on the hose, and the bellows expands, sealing off the pipe while emitting powerful, drain-clearing pulses of water.

TOTO LOW-FLUSH TOILET

It's inspiring that ecological awareness in the '60s and '70s caused toilets to be redesigned so that the typical 5 to 6 gallon flush was replaced by the 1.6 gallon flush. Trouble is, however admirable in water conservation, many of the early models didn't really get the job done, either requiring a plunger, or leaving what Robert Kourik refers to as "skid marks." Now, there is a new generation of low-flush toilets and some work better than others. One that works extremely well is the Toto 1.60 Gpf/6.0Lpf. In an independent ANSI test of 31 low-flush toilets, Toto scored highest for best performance—higher than more expensive models. Toto, in Morrow, GA, is the world's largest supplier of toilets.

SEPTIC PROTECTOR

http://www.septicprotector.com

Lint from washing machines is the leading source of solids that clog drainfields. Adding to the problem are fibers from synthetic clothing, which get into the drainfield, clog the pores of the soil, and do not break down as do natural fibers like cotton. This is a stainless steel filter for washing machines that keeps lint out of the soil; it's an excelwlent idea and a good product. It attaches to the outflow hose of the washing machine. It has a reusable 160-micron filter that you empty every few weeks. It comes with a 90-day, money-back guarantee.

GEOTECHNICAL GAUGE

W. F. McCollough, Beltsville, MD
301-572-5509

This is a unique, flexible, waterproof 5-inch-by-7-inch plastic card with nylon lanyard that contains miscellaneous data on characteristics of different soils. On one side are samples of six different types of common soil colors; on the other side are pasted-on actual sand grains, from coarse to silt. The sand grains and color chips are permanently fused to the plastic. Although this book does not go into soil analysis, this card can tell you what type of soil you have, and gives you field tests for determining clay consistency and type of sand in the soil.

DRIP IRRIGATION FOR GRAYWATER

AGWA Systems, Burbank, CA
818-562-1449 (fax), or
Real Goods Trading Co. 800-919-2400

Robert Kourik, author of *Drip Irrigation for Every Landscape and All Climates,* recommends the automatic graywater drip irrigation system manufactured by AGWA in his booklet, *Gray Water Use in the Landscape.* Mail order kits start at around $1000 and a custom system for a home can run between $3000 to $5000.

REAL GOODS TRADING CO.

Ukiah, CA
800-919-2400
http://www.realgoods.com/

The world's largest seller of renewable, eco-conscious products, Real Goods has composting toilets (Sun-Mar, CT Systems, Carousel, Biolet, and AE Hybrid), drip irrigation systems, water filters, and many other items that will be of interest to anyone dealing with septic systems.

absorption, 33
adsorption, 33
advanced systems, 86–99
  characteristics, 86–87
  mound systems, 89–90
  pressure-dosed drainfields, 88–89
  sand filter systems, 91–93
    components, 91
    reasons for, 91
    size of, 92
  wetlands, 98–99
    flow type, 98
    free-water surface type, 98
    how they work, 98–99
    purification by plants, 99
alternative systems for small towns, 127
  collection options for offsite treatment, 131
  common onsite systems, 128
  options for difficult sites, 129
  options for special situations, 130
Alth, Max and Charlotte, 168

bacteria, 31–32, 34
bathroom, modern, 150
Bio-Dynamic gardeners, 71
biofilters, 94
biological mat. *See* biomat.
biomat, 17–19, 32
Black Plague, 138, 140
blackwater, 64
Borregaard, Ebbe, 68
Bramah, Joseph, 145
Buer, M.C., 138
Burns, Max, 168

Cameron, Donald, 152–153
Center for Disease Control, 107
cesspits, 139
cesspools, 26–27
Chadwick, Allen, 71
chamberpots, 140
cholera, 140–141
Clean Water Act, 102, 126
composting toilet systems, 74–83
  batch processing, 76
  brand names, 76, 80–83
  build-it-yourself option, 77
  continuous processing, 76
  central composting toilets, 76
  faulty systems, 74–75
  maintenance, 77–79
    aeration, mixing and additives, 77
    carbon-to-nitrogen ratio, 78
    heat, 77
    moisture, 78
    monitoring pile, 78
    removing end-product, 78–79
  microflush toilets, 76
  self-contained composting toilets, 76
  site-built designs, 77
  vacuum toilets, 76
  why to install, 75
Controlled Energy Corporation, 40
Crapper, Thomas, 146, 170, 171
Crites, Ron, 126, 166
Cummings, Alexander, 145

da Vinci, Leonardo, 136
Del Porto, David, 83
DeVault, George, 108
diseases, waterborne, 31–33, 138–141
dishwashing system, 39–42
disposal field. *See* drainfield.
distribution box, 59
DNA testing, 105, 110, 112, 115
Drain King, 45, 57
drainage, 53, 158–161
  Intercept Trench System, 53
drainfield, 14–27
  construction, 14–15
  design, 15
  distribution systems, 16–18
  dual, 20–21, 53
  failure, 49, 58
  pressure-dosed, 22–23, 88–89
  shallow, 22–23, 96, 159
  valve, diversion, 20–21
Dunbar, W. P., 11
Duncan, Rick, 59

E. coli, 32, 110
E.P.A., 74, 102–104, 107, 109, 110, 112–114, 120, 127, 158–159
Ecos Catalog, 83
effluent, 6, 30–33
engineers, 120–125
  consulting, 120
  contract, 124–125
  interviewing, 123–124
  selecting, 124–125
evapotranspiration, 59

failure. *See* system failure.
Farmer's Home Administration, 120
Farwell, Larry, 69
filters
  effluent, 95
  Puraflo Peat®, 94
  trickling, 94
  Waterloo®, 94
Ford, Brian J., 34–35
Fosse-Mouras tank, 151–153
"French drain," 53

garderobes, 139
Gaspers, Michael, 70
Giardia, 31, 34
Giblin, Albert, 146
Goldberg, Rube, 77
gravel, 18
gravity, 18, 86
graywater systems, 64–71
  definition of, 64
  diverting graywater, 65–67
  dump truck drainfield, 70
  leach lines under raised bed, 71
  legality of, 65
  mini-drainfield, 69
  mini-septic system, 70
  small drainfield, 68
  surge tank for washing machine, 66
  valve, 3-way diverter, 67

Harrington, Sir John, 136, 144
Hellyer, Stevens, 147
history of wastewater disposal, 134–155
Hokusai, 143

household maintenance, daily, 38–45
  bath/shower water, 39
  dishwashing system, 39–42
  drain cleaners, 45
  garbage disposal, 42, 45
  grease and oil, 42, 45
  laundry water, 39
  leaky taps, 39
  paper products, 42
  sand and dirt, 42
  synthetic fibers, 42
  toilet, 39
  water conservation, 38–43
  water softeners, 45
Hugo, Victor, 74
Hulls, John, 112

Imhoff tank, 155
infiltration, 33
Infiltrator Chamber System, 22–23, 172
inspection, 49–53
  drainfield, 52–53
  tank, 49–52
Intercept Trench System, 53
irrigation, drip, 95–96

Jandy Industries, 67
Jennings, George, 134, 147

Kaplan, O. Benjamin, 167
Kley, Heinrich, 142
Koch, Dr. Robert, 141
Kourik, Robert, 170

Lambton, Lucinda, 171
Larson, Tom, 109
leachate, 75, 79
leachfield. *See* drainfield.
Lehman's Non-Electric Catalog, 83
loo, 140
Ludwig, Art, 65, 69, 169, 172

Maintenance.
  See septic system maintenance.
May, Randy, 18, 33, 104
microbes, 34
Mitchell, Michael D., 10
Moule, Henry, 148
mound systems, 89–90
Mouras, Louis M., 151–152

National Small Flows Clearinghouse,
    44, 90, 99, 109, 114, 122, 127, 171

Ochs, Ed, 108
Orenco Systems, 9, 21, 94–95, 172, 174
outhouse, American, 149, 171

pathogens, 34, 79
percolation, 33
Perkins, Richard J., 78, 167
pipe blockage, 56–57
Pirana/Sludgehammer, 97, 160
plumbing trap, 145
"Privie in Perfection,"144
pressure dosing, 22–23, 88–89
Pumper, 172
pumping, tank, 49, 52, 54
pumps
  maintenance, 54
  mound systems, 89
  power outage, 57, 89
  pressure-dosed drainfield, 88–89
  sand filter systems, 91–93

Queen Elizabeth, 144
Queen Victoria, 141

Real Goods Renewables Catalog, 83
"repair-as-is" option, 60
risers, 5, 49
  retrofitting, 50
roots, 53, 59
  blockage, 57, 59
RotoRooter, 57, 59
Rural Development Administration,
    122, 124

Salvato, Joseph A., 166
sand filters, 91–93
sand-lined trenches, 93
scum, 6, 51–52
  scum-measuring device, 51
seepage beds, 24–25
seepage pits, 25–26
septic system maintenance, 48–54
  drainfield inspection, 52–53
    drainfield test, 52–53
  locating tank, 49–50
  tank inspection, 50–52
    inlet chamber, 51
    outlet chamber, 52
    pumping, 49, 52, 54
sludge, 6, 51–52
  sludge-measuring device, 51
Small Flows, 44, 92, 99, 109, 122, 127
small town septic systems upgrades,
    118–127
  alternative systems for small towns
      and rural areas, 127–131
    collection options for offsite
        treatment, 131
    common onsite systems, 128
    options for difficult sites, 129
    options for special situations, 130
  special district, forming, 118
Snow, Dr. John, 141
soil absorption systems, 14–26
soil, 30–35, 158, 175
  absorption, 33
  adsorption, 33
  basics, 30
  infiltration, 33
  percolation, 33
  purifier, as, 33
  texture, 31
Steinfield, Carol, 74, 80, 83, 169
Stevens, Leonard, 35
system failure, 56–61
  drainfield failure, 49, 58, 60
    clogging with solids, 58
    distribution box solution, 59
    high ground water, 59
    root blockage, 59
  locating problem, 56–59
    flooding from inside, 57
    pipe blockage, 57
    power outage, 57
    root blockage, 57
    system blockage, 56–57
  tank failure, 58

tank, 4–11
  anaerobic decomposition in, 7–8
  construction, 4
  effective volume in, 7
  effluent filter, 8–9
  function, 5
  inlets, 8, 51
  inspection, 49–52
  meander tank, 10
  one-compartment vs.
      two-compartment, 5
  outlets, 8, 52
  primary treatment in, 6
  retention time, 7
  retrofitting, 50
  size, 4
  types of, 4, 58
  watertightness, 5
Tchobanoglous, George, 19, 99, 126, 154
toilets, 39, 64, 76
  Acme Water Closet, 150
  first flush toilets, 135
  Moule Earth Closet, 148
  Optimus Valve Closet, 147
  Short Hopper Closet, 146
  Syphonic Closet, 147
  Valve Closet, 148
  Washdown Closet, 148
  Washout Closet, 148
  See also composting toilet systems.
  See also "Privie in Perfection."
Trevelyan, G. M., 140

U.S. Public Health Service, 155

vadose zone, 23, 159
valve diverter for graywater, 67
viruses, 31, 32–34
Vogel, Nancy, 132

Warshall, Peter, 10, 30, 168
water conservation, 38–43
water seal, 145
wetlands, 98–99
Winneberger, Dr. John H. Timothy,
    10, 32, 44, 166
Wright, Lawrence, 137

Credits

Managing Editor
Robert Lewandowski

Contributing Editor (second edition)
John Hulls

Contributing Editors (original edition)
Blair Allen
Julie Jones

Production Manager
Rick Gordon

Design
Lloyd Kahn, Rick Gordon, Peter Aschwanden

New Cover Art Additions
Glen Strock

Design Consultant
Beverly Anderson

Scanning
Myrna Vladic

Proofreaders
Susan Friedland
Robert Grenier

Photo of Peter Aschwanden
Ada Browne

Printing
Courier Companies, Inc., Westford, MA

Special thanks to the following people for technical advice:
Steve Dix
Rick Duncan, R.S.
Larry Farwell
Brian Ford
Al B. Foreman
Daniel Friedman
Michael Gaspers
Beverly B. James, P.E.
Bonnie Jones
Graham Knowles
Harold Leverenz
Art Ludwig
Randy May, P.E.
Glenn Nelson
Orenco Systems
Michael D. Mitchell, P.E.
Durland H. Patterson, Jr.
Dale Rausch
Mark Richardson
Phil Smith
George Tchobanoglous, Ph.D.
David Venhuizen, P.E.
Tom Watson
Steve Wert, C.P.S.S.
John H. Timothy Winneberger, Ph.D.

Special thanks to the following people for help in various ways:
Kim Cooper
Alexander Walter

SOURCES AND ACKNOWLEDGMENTS

Chapter 3

Pages 34–35. Illustrations by Brian J. Ford from *Microbe Power,* Stein and Day: First Scarborough Books Edition, 1978

Chapter 13

In preparing Chapter 13, A Brief History of Wastewater Disposal, we are indebted to the excellent book, *Clean and Decent, The Fascinating History of the Bathroom & the Water Closet and of Sundry Habits, Fashions & Accessories of The Toilet, Principally in Great Britain, France and America*, by Lawrence Wright (The Viking Press, New York, 1960); unfortunately, the book is now out of print. It is a profusely illustrated and fascinating history of personal hygiene that is both scholarly and entertaining, and we hope to see it back in print before long.

Page 135. Redrawn from *The Palace of Minos,* Sir A. Evans. MacMillan, London, 1921

Page 136. Pipes, redrawn from *The Palace of Minos,* Sir A. Evans. MacMillan, London, 1921

Page 136. Seat, from *The Story of Baths and Bathing,* G. R. Scott. T. W. Laurie, London, 1939

Page 137. Redrawn from photo

Page 138. *Dictionnaire de L. Ameublement,* H. Havard. Paris, 1890–94. From *Aesop's Fables*

Page 139. *Principles and Practices of Plumbing,* S. S. Hellyer. Bell, London, 1891. Drawings by F. R. Dickinson

Page 140. "La calavera catrina," José Guadalupe Posada, from *Posada's Popular Mexican Prints,* Roberto Berdecio & Stanley Applebaum. Dover Publications, New York, 1972

Page 142. *The Drawings of Heinrich Kley.* Dover Publications, New York, 1961

Page 144. Reprinted in *Flushed With Pride—The Story of Thomas Crapper,* Wallace Reyburn. Pavilion Books Limited, London, 1969

Page 145. Both drawings from *The Plumber and Sanitary Houses,* S. S. Hellyer. Batsford, London, 1877

Page 146. Hopper closet redrawn from promotional literature of George Jennings Esq., London, 1858–1900

Page 148, top. Redrawn from *Sanitation: An Historical Survey.* The Architects' Journal. London, 1937

Page 148, bottom. John Bolding & Sons Ltd.

Page 151. Redrawn from *Principles of Sewage Treatment,* W. P. Dunbar. Charles Griffin & Co., Ltd., London, 1908

Page 152. Same as above

Page 153. Redrawn from *The Antecedents of the Septic Tank,* Trans. Am. Soc Civ. Eng. XLVI, 1901

Page 154. Redrawn from *Small and Decentralized Wastewater Management Systems,* Ron Crites and George Tchobanoglous. McGraw-Hill, New York, 1998

Glossary

Most of the glossary definitions are from *Glossary of Terms in the Wastewater Industry,* by California Association of Sanitation Agencies, edited by Harriette Heibel, published by the Central Contra Costa Sanitary District, Martinez, CA, 1989.

FOR SMALL COMMUNITIES AND WASTEWATER DISTRICTS

To distribute to local homeowners

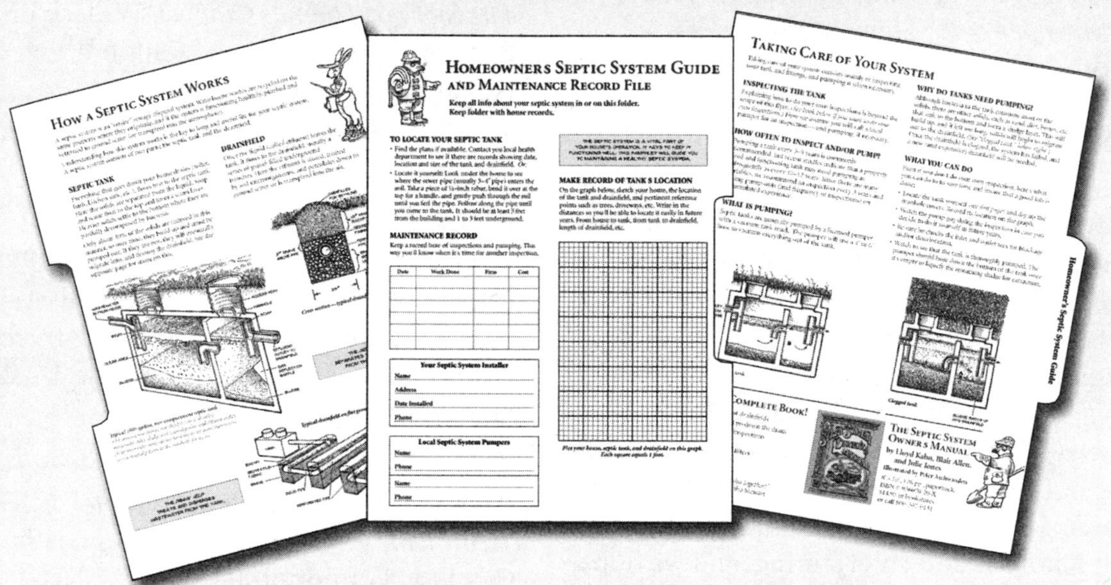

Homeowner's Septic System Guide and Maintenance Record File

A Low-Cost Guide and Maintenance Record in a File Folder

A place for homeowners to:

- Show tank location in relation to house
- Keep phone numbers of local pumpers
- Record dates tank is pumped
- Keep records on hand and pass along to new owners if house is sold

Includes as well:

- Description of tank and drainfield
- Basic maintenance instructions
- Tips for homeowners

Available in quantities of 10 or more

Contact Shelter Publications, P.O. Box 279, Bolinas, CA, USA

800-307-0131 (toll-free)   415-868-9053 (fax)   orders@shelterpub.com

For Online Orders
SHELTER ONLINE
http://www.shelterpub.com

The 1973 Classic

Shelter

edited by Lloyd Kahn

$24.95
11″ × 14½″
176 pages
ISBN-10: 0-936070-11-0
250,000 copies sold

"How very fine it is to leaf through a 176-page book on architecture—and find no palaces, no pyramids or temples, no cathedrals, skyscrapers, Kremlins or Pentagons in sight . . . instead, a book of homes, habitations for human beings in all their infinite variety."
–Edward Abbey, *Natural History* magazine

"A piece of environmental drama."
–*Building Design* magazine

"An embarrassment of riches."
–*Manas*

With over 1,000 photographs, *Shelter* is a classic celebrating the imagination, resourcefulness, and exuberance of human habitat. First published in 1973, it is not only a record of the countercultural builders of the '60s, but also of buildings all over the world. There is a history of shelter and the evolution of building types: tents, yurts, timber buildings, barns, small homes, and domes. There are sections on building materials, heavy timber construction, stud framing, stone, straw bale construction, adobe, plaster, and bamboo. There are interviews with builders and tips on recycled materials and wrecking. The spirit of the '60s counterculture is evident throughout the book, and the emphasis is on creating your own shelter (or space) with your own hands. A joyful, inspiring book.

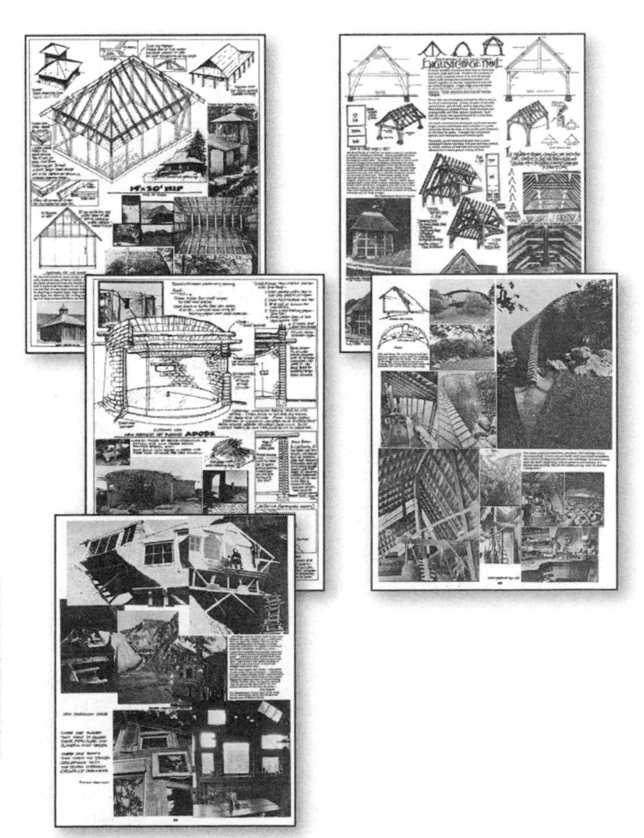

Shelter Publications
P. O. Box 279
Bolinas, CA 94924 U.S.A.
415-868-0280 1-800-307-0131 (orders)
orders@shelterpub.com www.shelterpub.com

The Sequel to *Shelter*

Home Work

Handbuilt Shelter
by Lloyd Kahn

$26.95
9″ × 12″
256 pages
ISBN-10: 0-936070-33-1
ISBN-13: 978-0-936070-33-9

"The book is delicious, soulful, elating, inspiring, courageous, compassionate . . ."
–Peter Nabokov, Chair, Dept. of World Arts and Cultures, UCLA

"Home Work *is a KNOCKOUT."*
–John van der Zee, author of *Agony in the Garden*

". . . magnificent, invigorating, deep . . . and simply inspiring."
–Kevin Kelly, author; former executive editor, *Wired* magazine

Home Work is Lloyd Kahn's latest and most ambitious book, a stunning sequel to *Shelter* that illustrates new and even more imaginative ways to put a roof over your head, some of which were inspired by *Shelter* itself. *Home Work* showcases the ultimate in human ingenuity, building construction and eco-centric lifestyle. What *Shelter* was to '60s counterculture, *Home Work* is to the Green Revolution, and more.

Home Work describes homes built from the soul, inventiveness free from social constraint, but created with a solid understanding of natural materials, structure, and aesthetics. Yurts, caves, tree houses, tents, thatched houses, glass houses, nomadic homes, and riverboats—each handbuilt dwelling finds itself at one with its environment, blending harmoniously with the earth, using organically sustainable material.

Home Work features over 1,000 photos, including 300 line drawings, stories of real people building and living in their own houses, plus Kahn's recollections, reminiscences and observations gathered over the 30 years since *Shelter* was first published.

LOUIE FRAZIER

THE YURTS OF BILL COPERTHWAITE

TIMOLANDIA